ATTILA PERGEL

THE BOOK THAT HAPPENED

IS REALITY BUT SHEER COINCIDENCE?

Author: Attila Pergel, pergel@pergel.hu

Editor: Krisz Nádasi, www.krisznadasiwrites.hu

Copy Edit: Judith Henstra PhD - Book Helpline - www.bookhelpline.com

Translation: Abigail Péter, abipeter.95@gmail.com

Illustration and Cover Design by Klára Somogyi, klarasomogyi12@gmail.com

Profile picture: Szabina Hauberger, szabinahauberger@gmail.com

First Printing, 2021

DEDICATION

To my son, Dávid, even though he's not old enough to contemplate the questions the book offers. However, the book's main message is that time is only a matter of perspective—so, even if the book has a perpetual message, the dedication comes true whenever he decides to pick it up and starts reading.

Did You Find This Book by Sheer Coincidence?

Why was it my car that was totaled in an accident? Why was it me who won the card game, and why was it my opponent who lost? And how about that sheer coincidence when I met my wonderful wife? What was the event that set the writing of *The Book that Happened* in motion? If we contemplate our life events, we'll find not one or two, but billions of coincidences without which our life would have taken a different course. On any given day, numerous connections can be found that we prefer to see as pure chance.

But does coincidence really exist?

In this book I deal with this question. I'll show connections that help you understand the reasons of coincidence and that it is a matter of perspective. I'll also point out how coincidence often only appears to be that, and instead is fake. You might not have seen it coming, but somebody else did. Let's put our heads together, read the pages, and do some mental experiments. I promise that it will be a fantastic adventure!

When we reach the end of the book, we'll know how to decide if an event was caused by coincidence. We'll find its true meaning, even though we'll begin with a clichéd definition. It will be inevitable to touch infinity, travel through dimensions and multi-dimensions, admire quanta, and the genesis of our universe. We'll also deal with the concept of time, in a way that may surprise you.

The book offers a way to catch coincidence in action. It provides a method to understand complex coincidences quickly and to picture them logically and physically. It shows a simple way to compare groups of coincidences. Where we dig a little deeper into physics, I'll give clear and simple explanations. I strive to provide food for thought without demanding a complex educational background, so you only need to be able to think logically.

How come I know so much about coincidence? I wrote my very first computer program at the age of thirteen; it was a dice program. I did that back in 1985 on a ZX81 Sinclair home computer. I've been a software developer for more than thirty years, and in the past ten years I've been working for financial institutions. Programming makes use of coincidence in many areas—or to be more specific, programming uses artificial coincidence for calculations. For example the secure use of credit cards was made to happen by creating artificial coincidence — this is what we call encryption. This means that the economy of today—the entire global economy—is based on computer science, on cryptographic systems. If the job is done well, the bad guys are unable get hold of data that can be used to harm us or others.

Quite amazing, isn't it? How could the security of the global economy rely on coincidence?

Ever since I was a child, I've been eager to find out the truth about coincidence. My curiosity grew into annoyance in the last couple of years. I still knew so little about coincidence, and yet I used it on a daily basis.

People say that everything happens for a reason; nothing happens by chance.

But do they really mean it?

I know I didn't.

But then I took Bell's theorem and carried out an experiment via a computer program. There was an error in the code, and the chart result showed an extraordinary image. I couldn't believe that in experimenting with coincidence I could accidentally make such a mistake. After this, I felt great respect for coincidence, and I finally got to the starting point of my discovery.

The book, furthermore, aims to grasp the meaning of Hermes Trismegistus's duality that he stated as follows: *"Every nature which lies underneath should be co-ordered and fulfilled by those that lie above."* He says that *underneath* and *above* only make sense together: one cannot exist without the other. We'll see.

It doesn't matter if you accept or disagree with the questions and answers of the book; they are meant to help you better understand your life and your world. Having read this book, you may find to be true what you always considered to be false and vice versa. It will be up to you to pick a convention you agree with and judge what is true and what is false—you'll see that it's all arbitrary.

The Book that Happened is not here to tell you that you are wrong. I simply want to show you why I think I'm right and how I found my way to my truth. I fully reserve the right to have a truth that doesn't necessarily corresponds with yours. I am also fully aware of the fact that over time, the truth I talk about in this book may lose its relevance or may even become part of someone else's truth.

Table of Contents

I.

BUILDING THE REALITY

1. What Makes up a Coincidence?

The present chapter analyses seemingly basic notions and phenomena such as time or light. This, however, won't be as simple as it sounds. In my experience, the more complicated a system really is, the simpler it looks. If that statement is true, then so is its opposite: There is a simple system behind every complex-looking thing. Let's take a blade of grass. How simple it seems! But if you think twice and consider what you know about it, you'll be amazed.

A blade of grass can turn inorganic matter into organic matter.

We, who consider ourselves to be superior beings to a blade of grass, are not so good at that.

We know that full quantum mechanical processes take place in a blade of grass. We even named it—photosynthesis—yet we have only just begun to understand it.

If I want to address the idea of complexity by comparing a Boeing 747 to a blade of grass, then the blade of grass is a supercomputer, and the airplane is a dowel. Take the structure of the Rotating Tower in Dubai. It looks quite complicated—it truly is remarkable. But a closer look will prove that the materials that were used in its construction are anything but complicated: concrete, glass, iron, and steel. They were put together according to a clear and comprehensive plan. The Rotating Tower, therefore, is not complicated, because we can see what it is made of and how the materials are connected to each other.

But what should we consider complex? What should we consider simple? Fair question. These words are indeed arbitrary.[1]

Still, let's try and define complexity.

We may say that something is not complex if we understand its structure, origin, and mechanisms. Let's take a look at the objects we are surrounded by. They are made by men, usually after precise planning, knowing exactly what to expect from a particular material in the end product. Therefore the makers of artifacts are aware of all circumstances of the creation of these artifacts, and they have all the necessary plans, so it is safe to say that we people are not able to create complicated structures. Complexity depends on the person looking at it: Does he find it easy or difficult to understand?

We don't know how natural creation works though. We have no knowledge about the methods and mechanisms that are being used, and we only speculate about the materials too. That means that whatever nature creates is complex by definition.

---

[1] The mathematician László Lovász stated, "There is no algorithm to calculate the complexity of a particular sequence."

It's time to put fixed ideas, stereotypes, and preconceptions out of the way. As you go on reading, you'll find that this isn't an easy task. We've been letting our brains be programmed since birth, and everything that now fills our brains seem so obvious that we forget to question the reasons and goals of things happening around us.

Let's try a quick and simple experiment, and we'll see how effective the programming has been.

Think about three relevant events in your life. Choose one that happened ten years ago, another that happened five years ago, and a third, recent one from these past weeks or days. Create three separate mental pictures.

Ready?

Now put them next to each other: first the event from ten years ago, then the one from five years ago, and then the most recent one. It all looks nice together, right? But let's now switch on our defense system by changing the sequence of the pictures. See if your brain lets you put the most recent memory first, replacing the oldest one? Then the second picture stays the same, and the third picture will be from the oldest event.

As I said, this is a simple experiment. But don't feel discouraged if the reorganizing does not happen quickly or doesn't seem to work out on the first try. Maybe you get there after three to five attempts.

This is how many it takes to trick the unit that is responsible for keeping the flow of time intact. And if you were able to do it on your first attempt, now try putting four or five mental pictures in reverse order.

This simple test points out how our minds accept doctrines to an extent that it fights back when we happen to disagree: *What are you thinking? Time can only head in one direction!*

Well, this is exactly the way of thinking that we have to forego to become more open-minded. The good news is that the brain is open to training. As we go on reading, it will be easier and easier to visualize ideas that hardly ever cross our minds in the daily hustle and bustle of work, shopping, and doing household chores.

The first part of the book is about those so-called facts and phenomena that we simply take for granted and don't bother questioning. According to our brain, these are basic concepts: *The answer is so simple, accept the facts.*

So *The Book that Happened* begins with these concepts. We lay down the foundation on which we start building. We look at seemingly simple things, only to discover how unthinkably complex they truly are. I urge you to be on the lookout for contradictions. Don't be contented with my explanations; don't take everything at face value. Question me! Judge everything with a stern eye, and use logical thinking; try and find fault in whatever I say. The more questions you have, the more you get to understand, and the more answers your brain will supply.

You may first ask: why logic? Plato said that once science contradicts itself, a new theory has to be developed, one that would solve the contradiction. He also considered that the starting point of the new must be the old. But what did he mean?

It's not easy to question Plato's words. They can be true and false at the same time. Just as a knife can be used for a good deed—slicing bread—and evil—murder. If we use Plato's theory to confirm a result within our own system by all means—because it generates adequate results and enables our model to function—it may prove destructive. When we want to show that we are right, we should never use theories that confirm our statement. Instead we should look for theorems that seem to contradict us and see what's behind them. We might find that they are not true—or that we were wrong.

The ancient Greek physicist, mathematician, and astronomer Aristarchus of Samos (c. 300 BC) had to suffer the destructive power of false theories. He was the first to state that the sun and the stars are fixed and that it is the earth that revolves around the sun. Religion, however, followed Plato's theory, considering that the starting point of the new must be the old. Since they didn't find any contradiction between the geocentric worldview and the holy texts, they did not see any reason to change their beliefs where the sun and the earth were concerned. Aristarchus of Samos, however, considered the theories of the holy texts incompatible with his own. He became the first scholar to devise the model of heliocentrism. Aristarchus was accused of heresy by Cleanthes, and ultimately had to flee.

For two thousand years, Plato's theory was a matter of life or death. Heliocentrism played a part in the death of Hypathia, the Alexandrine scholar. Then in 1600, we also punished Giordano Bruno for his heliocentric view: We burned him alive. Today's astronomy would not be the same without the contribution of Galileo Galilei, yet we dragged him through the mud too. He narrowly escaped execution by burning at the stake, and he had to endure long years of imprisonment, followed by house arrest, and he was prohibited to write. Galileo Galilei reconciled the contradiction in his own system, but it nearly cost him his life.

As we see, the clergy at the time evidently justified themselves through their own texts and theories. If they happened to find contradictory information, they resolved it with a simple stroke of pen. By such corrections they fulfilled Plato's requirements, having created their own science.

The great truths of the past can only be considered true if they are interpreted correctly. If we keep on building on a false foundation, we won't reach truth in the end. Especially if we do it in cycles, we'll certainly reach lies. Therefore, we must go back to the start, demolish the foundation, and lay down a new and stable one. It will be a tough job! We might not like the results of some logical deductions and mental experiments. Others will seem more likable. But results don't care about being liked or disliked. They won't be happier if they're readily accepted, and they won't feel awkward if they're rejected because they were too hard to understand.

Mental experiments are crucial in this book. They build upon logic and yield results. Most experiments aim to tackle a proven concept, and they do so through simulation. Simulation will often differ from the real way that leads to the result, but since we already know what result to expect, we will declare the simulation true. In such mental experiments we can exclude unnecessary factors that would only complicate the process. We will also have the possibility to include hypothetical elements that will lead us to desire the result, therefore proving the hypothesis correct.

The following chapters focus on factors that play a part in coincidence. We need to understand them first in order to get to further, abstract definitions that cannot be explained rationally.

1.1 Time

Have you ever wondered about the concept of time as we know it today?

I suppose there were situations when more people had to do something—go hunting or perform a ritual sacrifice, for example—long before the introduction of accurate time tracking. They must have had mutual agreements to time these events, maybe the position of the sun. One thing is sure: At some point someone discovered that if they stuck a piece of wood in the ground, its shadow would reflect the movement of the sun. And the exact same thing would happen the next day, too. Maybe he realized that by placing rocks on the route of the stick's shadow when he went hunting or when the ritual sacrifice was held, he could be sure that when the shadow of the stick reached that particular rock the next day, it would either be time to go hunting or time to attend the sacrifice.

This must have become overwhelming for our inventor after a while, since he must have kept adding rocks around the shadow of the stick. Each of them was linked to a particular event, but they appeared in a random order. The fact that hunting or ritual sacrifice were not an everyday event made matters even worse. The shadow touched each rock day after day, no matter which of the events were supposed to happen every day.

Despite the random placement of the rocks, our inventor must have been able to see that the rocks formed a certain shape around the stick—an oval. It might have been this discovery that led to the invention of the sundial.

So we can make a statement regarding how we started measuring time—our ancestor measured events instead. A chain of events, made up of millions and billions of movements. As the shadow moved, he placed event timers connected to particular happenings. To put it plainly: He counted and measured movement.

Our inventor could have been sure that when the shadow of the stick reached one of the rocks, it meant that something was about to happen. For example, it was time for the sacrifice ritual. But what if the earth started spinning faster for some reason? Say, for the following three days, it revolved 30 percent faster? (Let's omit natural and ecological disasters, but suppose that speeding could happen without serious consequences.) What would our inventor have noticed?

Well, nothing at all. The shadow of the stick would still have reached the rocks the same way as before. In the evenings, our inventor might have felt exhausted since the day was shorter and he still had to perform the same amount of tasks. He had a shorter timeframe for all that—because of faster the rotation of the earth.

From a scientific perspective, it is evident that our inventor picked a wrong reference, because if the speed of the earth's rotation changed, he would not know that his timing was incorrect: He wouldn't be able to perform his tasks at the exact time he planned. If the very foundation is unreliable, there can be changes that are impossible to detect—the shadow of the stick would still move from rock to rock.

Here's another question worth pondering. Let's disregard the moon phases and the orbit of the earth around the sun for a second and imagine that our planet doubles its speed as of tomorrow without any serious consequences. How would it change the average life expectancy, which today is around seventy-five years? Starting from tomorrow, would it be seventy-five or thirty-seven years? (We will come back to the issue of years, months, and weeks later, for no other reason than to point out how shaky is the ground on which we have built our knowledge.)

Thanks to the development of technology, the hourglass was invented. Like the sundial, it had its flaws. If the sand was too dry or too moist, its speed changed accordingly. People were right to ask, *Why does time slow down when the air is damp?* The sand did flow at a slower pace. The hourglass therefore measures movement, not time. Again. Certain external factors can change the frequency of cycles. Therefore, the hourglass depends on the environment, and therefore, we identified yet another unreliable instrument for measuring time.

In the past, we have used all kinds of mechanical instruments to measure time. The most recent one is the atomic clock. To put it simply, the clock uses "the electromagnetic spectrum of atoms as a frequency standard for its timekeeping element."[2] The frequency is presented as seconds (and larger units of time) on the display unit of the clock. The operators of these clocks do everything in their power to keep the electromagnetic spectrum unchanged and undisturbed by the various external conditions, I am sure. Nevertheless, the frequency depends on the environment: temperature and the electromagnetic field. There are solutions to these problems, to keep the frequency unchanged. Up to now, it seems that atomic clocks work accurately, and the atomic frequency shows the same values in any given environment—but maybe it just *seems* so. Do think about it. In prehistoric times, the sun came up each morning and it set each evening, for days on end. Imagine the surprise and confusion when there was a solar eclipse! They happen only every seventy to eighty years. We, with our atomic clock, also suppose that the variables in nature are constant. To define these variables, we use measurements that are based on other unstable units of measurement.

The way we measure length and weight is a perfect example of that. Both units of measurement are based on mutual agreement and depend on external factors. Take for example a 1-meter-long piece of steel. The statement that it's 1 meter long is true only when the temperature is 20 Celsius (i.e. 68 Fahrenheit). As soon as the temperature rises, for example to 104 Fahrenheit (40 Celsius), the length of that same piece of steel will *not* be 1 meter anymore.

---

[2] Wikipedia article about the atomic clock.

The situation is the same with weight: It depends on *where* we measure. If you measure 1 kilogram on the seaside, and then take it to 3000 ft above sea level, it won't be 1 kilogram anymore. And if you take it to the moon, you get yet another completely different result.

And it's the same with similar units, for example foot or pound: They are based on agreements and variables.

Therefore it is safe to say that atomic clocks are only as accurate as the variables of the environment are constant.

One more thing about the atomic clock: What if there is a global mistake in the calculations that control the environmental variables? It could hide the problem itself, the same way as our inventor wouldn't know the Earth revolves faster just by looking at his sundial. Would we notice anything? I think not. So what we can state about the atomic clock is that its foundation is based on a variable, which variable was defined by calculations and measurements that themselves depend on other factors.

Humans aim to know and prove what the length of one day is, i.e. the time needed for the earth to turn around once. The introduction of *Coordinated Universal Time* (UTC)[3] looked promising. To compensate for the slowing pace of the earth's rotation, UTC ad-

---

[3] Wikipedia article about UTC.

justs the length of a year by adding roughly one second to it, based on the measurements of the atomic clock. This is the only way to get the *real* length of a day, which is considered to be 86,400 seconds.

If we project these leap seconds of UTC to the world population of eight billion people, we suddenly get a bit more than 253 years out of nowhere. And this is only the smaller part of the problem. Much more disturbing is that there is no instrument or logic to measure time which doesn't depend on the environment, which isn't relative. Just think about UTC. It is built on two inaccurate measurements—the rotation of the earth and the calculations of the atomic clock. We now believe that the atomic clock is more accurate than all other means we've tried before, therefore we've made it the standard, and we're adjusting the time we calculated based on the earth's rotation with the atomic clock's measurements.

All of the above might lead us to the conclusion that time is what we experience between two events or movements. Keep this in mind, because we will see that time flies differently for an atom or a human being, as soon as it is taken out of its natural environment, the frame of reference is omitted and the environmental variables are changed.

The earth does not only revolve around its axis, but it orbits the sun as well. Back in the time of the sundial, our inventor must have noticed that with the passing of many, many days, the shape drawn by the stick's shadow changed a bit. He must have discovered that some events happened periodically, following a cycle: snowfall, for example, or blazing sunlight. He must have observed that there was a main cycle every 365 days. He called it a year. But this created another problem: He had to divide the year into shorter sections to be able to place the periodical events. But how could 365 days be divided into equal periods?

There must have been a huge discussion about what's to be done. Saying, *"See you in 143 days"* or, *"Come over in 287 days"* to a friend is quite impossible, isn't it? Perhaps our ancestors were influenced by the planets. Perhaps some group pointed at a number they liked best. We will never know, but one way or another, they came up with the number 12. (I have given it so much thought. Why not *ten*?) This number maybe reflected the phases of the Moon, or the geometric circle, which is 360 degrees, and therefore can be divided by 12. But this number doesn't solve the problem. Maybe it was used because ancient Egyptians defined a 360-day structure for a year—based on how many days passed between two floods of the Nile (it was between 360-370 days back then).

This is logical: 360 days can actually be divided equally into 30 days per month. But the introduction of the Julian and Gregorian calendar changed everything. The previous year of 360–370 days was suddenly replaced by another truth based on a new observation. It still used the 12-month structure, but raised the days of the year to 365 plus a little. And there were two problems with it. First, 365 cannot be divided by 12 without a remainder. Second, there's no such thing as that "little," i.e., a fraction of a day. A way had to be found to squeeze those unstable days into an unstable year in a way that creates a reasonable system of months and days. Maybe they flipped a coin or something. But in the end they chipped away some days from one month, and added some to another, without any apparent reason. And this is how our months came to exist.

The number of days in a month raises other problems of the system. For example, in Hungary people receive monthly paychecks, that is, the same amount of money each month. Here's a rhetorical question: Why is Hungarian labor worth more in February, which has twenty-eight days, than in March, which has thirty-one days? Which paycheck is real? The monthly or the daily? And there's another issue concerning finances: If we look at our bank statement of the previous leap year, is the yearly interest calculated upon 365 or 366 days?

But back to the creation of the calendar. The problem of weeks needed solving, because a month was still too broad when it came to dating an event. A new element was devised: calendar weeks. The length of the week can be tied to the division of lunar months into two or four periods. It shouldn't come as a surprise at this point, but this doesn't lead to exact numbers either, because four times seven is twenty-eight. Just a minor detail, nothing to worry about. By this point we're quite used to it, right?

However there were some merchant communities where one week was made up of only five days, and there were others who used eight. Some say that in the beginning weeks weren't even intended to be connected to months. And yet the calendar week became almost inherent to our mundane existence; we build on it even though we find no reasonable, scientific, or logical explanation as to how and why it exists in this form.

Anyway, it is obvious that making weeks compatible with months wasn't exactly a walk in the park, so the mission was abandoned altogether. It would have meant January ending on a Thursday, immediately followed by a Monday, being the first of February. (Or a Sunday, which is considered the first day of the week in several cultures.) This solution would've been hard to systemise, so the days of the week stayed continuous.

We see the fearlessness in the definition of weeks and months and years. The fact that the entire system was haphazard didn't seem to bother the decision makers, and neither did the lack of unity or accuracy. They seemed to have one goal alone: to give the people weeks and months and years. This sounds as if I sold a finished, yet defective software to a client, where for example logging into the program was only possible on every fourth attempt. If the client complained, I would wisely answer, "I did this for your own good, you see. The program is actually a lot safer, because nobody will have the patience to hack into it. So this isn't a bug, it's actually an advantage!" And in a way, I would be right, because truth is a matter of perspective. It all depends on which truth we pick: mine or my client's?

We can conclude that the calendar week lacks all sorts of design and accuracy. It was pressed upon us; we are forced to live by it, working for five days and resting for two.

Time wouldn't be complete without days, weeks, months, and years. We defined 86,400 seconds as one day, knowing it depends on the environment and various reference points, and further divided it into milliseconds, microseconds, nanoseconds, and so on.[4] The fact that we used the decimal system (where the units are powers of 10) for time units shorter than one second, is quite intriguing, because this too is arbitrary. Anything less than a second is actually so small that calculations that involve switching between the decimal and the sexagesimal system become extremely complicated and time-consuming. On the one hand, we would have to work with an infinite number of decimals, and on the other hand, the calculations would mostly be about conversion from one system into the other. It's quite strange that we hardly ever care about the

---

[4] Fun fact: 86,400 divided by 360 (a geometrical circle is 360 degrees) is 240, which is the sexagesimal numeric system itself.

exact number of seconds per day, but once we get to microlevel it matters when exactly an event happens within the second.

The evolution of time is remarkable. Units of time were continuously built on errors. We believed that the movement of the sun, the moon, and the earth is constant, and we used this as a point of reference for every calculation and theory of time. There have been attempts to repair the inherent mistakes, such as the introduction of leap years, leap seconds, and UTC itself. But I firmly believe that a house can only be well-built if the foundation is strong enough. And when it comes to time, we can hardly even talk about a foundation because of the millions of environmental variables.

We can analyze time logically instead of in a physical way. We are still bound to stumble upon contradictions that don't have rational explanation—and yet, time exists nevertheless.

Let's ponder the question of what is time. What do we mean when we talk about time in general? Does the past, present, and future actually exist, and if so, how can we logically prove it?

There's a paradox I like very much. I might have come up with it myself, but it's also possible that I've heard it somewhere. It goes like this: *If you do something today, how will you prove tomorrow that you did it yesterday?* The reason why we cannot answer correctly is because we need to involve the concept of *past* in the answer. But if we mention the word "past", we have to explain it, too, and we cannot do so without repeating it.

It is generally thought that the earth was formed 4.5 billion years ago. So here's another question: How did we get to this number? There was only one way: by relying on the structure of time that is being used today. And this system is based on comparisons.

However, we don't know what a day actually looked like four billion years ago. How may units made up the thing that we call one day, today? What today is approximately 86,400 seconds and a little, may have been only 40,000 seconds back then. The frame of reference for time measurement—the sun and the earth—probably didn't even exist at that point. We can only state that the earth was formed 4.5 billion years ago *according to how we measure time today*, but in reality, it could've been 1 billion year or 8 billion years.

Carbon dating is just as relative. We suppose that the decomposition of isotopes happened in the exact same way, with the exact same speed 40,000 years ago as it does today. With carbon dating, we can go back 50–60 thousand years in time, although the results are unreliable. The accuracy limit is considered to be around 37,000 years. I could say the very same about carbon dating and calculating the age of the earth: They are true only if everything that happens today, happened the same way yesterday, the day before, and 30,000 years ago, and even two billion years ago. So it would be true if we lived in a static world that never changed—but our environment changes at an unthinkable speed. As we examined the calculation of the earth's age, it became clear that we were dealing with something *intangible* and *impossible to uphold*. That's nasty. But unfortunately, time is a key factor of speed, so the miscalculations of time have a direct effect on speed—if one is flawed, so is the other. If we say that a day was made up of 86,400 seconds four billion years ago, then that would mean that the rotation speed at the Equator had to be 1,040.421 miles per hour. However, we know that this wasn't the case. The rotation speed was higher;

therefore, a day could not have been made up of 86,400 seconds for sure. And so, there's no point in taking 86,400-second calendar days into consideration when we want to talk about time in the past, and we can't talk about speed either. Whatever we say wouldn't be true because it would be calculated upon non-true time.

We can look at time from the perspective of biology. Who would know more about the passing of time than humankind? Aren't we the ones who are subjected to it, who live in it, experience it?

Let's see how time works when we are not awake. Have you ever experienced the presence of time in your dreams? Have you ever seen a clock tower or a wristwatch or any sort of clock? Have you ever dreamed that you had to finish something in time, or you were expecting something to happen at a given moment? Highly unlikely. Time hardly ever appears in our dreams; it is not connected to our actions. In our dreams, time usually stands still, and events *just happen.* We can't say that we experience time while we are dreaming, except, perhaps, if we refer back to how time and speed are linked together. In that case, there is some sort of experience of time since we do encounter speed in our dreams. Just think about those times you dreamed that you were falling into the abyss.

If time really was as natural as we suppose, shouldn't it be somehow displayed at a genetic level too? If we wake up in the middle of the night for some reason, we're hardly able to tell what time of day it is, let alone the exact hour. Have you ever woken up after a 5–10-minute nap, feeling as if you slept for at least an hour? Or the other way round: an hour-long nap felt like a couple of minutes only? I'm sure you know what I'm talking about. It seems that our brain isn't wired to know time; for some reason, it doesn't need it constantly.

So when does time—which otherwise seems so essential—become unnecessary? Let's take a look at how our brain works, and we'll find out. Our neural activity is based on electromechanics. And where there's electricity, there are electromagnetic waves too. Our brain produces such waves, that we categorize into four main states, based on our present knowledge and measuring technology. (Waves and frequency will be addressed in more detail in *Chapter 1.3 Light.*) In two of these categories we are able to discern time, while in the other two time is absent. It's important to note that the frequencies below are not the reason we experience time; they are physiological phases. Whether we perceive time during these frequencies is their consequence.

- *Beta state—13 to 40 HZ (13 to 40 cycles/second)* covers most of the period when we are wide awake. This is the state when we work, socialize, run to catch the bus, etc. This is the state that we call life, what we experience the most. In beta state we are highly aware of time; some people are able to track it with minute precision without checking a clock.

- *Alpha state—emission of 8 to 13 HZ electromagnetic waves.* It is a state when we are relaxed and at ease, yet our brain stays active. When we are resting or relaxing, we reach an alpha state; it is essential for daydreaming or artistic creativity. Intuition arrives in this state. However, the alpha state is not connected to physical activity or the lack thereof. We might also reach this state during a walk in nature, when we notice the beautiful landscape, or during yoga exercises (even though some of those require quite the physical effort). In this state, we can lose sense of time; we'll feel it was shorter than the actual measured period—this is when we feel that time just *flew*. The longer we stay in this state, avoiding the beta state, the less accurate we get in guessing time.

- *Theta state—emission of 4 to 8 HZ electromagnetic waves.* The higher frequencies of theta state make us slow and languid; the lower frequencies evoke images and fantasies. We often feel we control these events, but usually our consciousness flows with these flashing ideas. In the lower frequencies, we use our sense of time. By learning how to mix theta and alpha states, we are able to reach deep meditation, where we perceive an hour as five or six minutes. This explains why yogis and experienced practitioners of meditation can sit still for eight to ten hours. They feel as if only one hour has passed, or maybe even less, because they are able to arrange theta and alpha cycles consciously. If we are able to guess what time it is when our sleep gets interrupted, it means that we were in theta state only for a short while. During sleep this state brings dreams; in theta state, there are no paradoxes; everything seems obvious. Reaching a theta state while being awake enables us to recall long-forgotten memories and to creativity without limitations. Extraordinary ideas are born on the threshold between alpha and theta state.

- *Delta state—our brain activity is between 1.5 and 4 HZ.* Delta is the state of the unconscious, where time ceases to exist altogether. After waking in this state, we wouldn't remember that we woke up, we go back to sleep and reach delta state again at once. This state enables bodily functions that are inactive during the day, such as self-healing. For most people, it is impossible to reach delta state while awake, as if there was a veil between beta and delta state that is impossible to see through from either side. In delta state, time loses its meaning; reaching this state while awake is the deep trance. Some of the few who managed to experience the timelessness of delta state were Buddhist monks, creating a delta state by their so-called self-mummification. The mummification process takes three thousand-day long phases, the last one being spent in delta state. What in "reality" was several hundred days, felt for them only a few days.

Let's take another example that proves that time is relative: the concept of *dog years*. It's a curious little invention, to enable comparison between the age of dogs and humans. One dog year is said to be the equivalent of approximately seven human years. (There is some disagreement on the exact ratio, but seven is somewhere in the middle.) The comparison is a good idea as it gives us the opportunity to investigate a different time experience. Assuming that a dog experiences its life as long as we, humans, consider ours, it raises a few questions.

To enable a dog to experience seven years' worth of time in one year, it seems logical that it would have to experience seven times as many events. This is because both physics and biology measure time by using the cyclicality of events. We know that dogs spend half of their lifetime asleep, while for humans, it's only one-third. This is important because we don't perceive events and therefore time when we are asleep. The chart below shows how many events a dog should experience in order to match the experience of a human (i.e. events and movements).

	Human	Dog
Average life expectancy (years)	76	10
Years spent asleep	25	5
Years spent being aware of time	51	5

The chart makes it clear that a dog should experience ten years' worth of events in one year, i.e. ten times more events every day than a human. This statement transports us into the idea of dimensions; however, let's suppose that a dog has the exact same experience of time as humans. Then we must conclude that a dog experiences every event in much fuller details than humans. To be more exact, ten times fuller. This also means that each event feels ten times longer for a dog. Simply put, one minute for us is ten minutes for a dog.

All of these might seem strange, but this is where logic leads us. And we can formulate this idea by saying that when we leave our pet home alone for twelve hours, the dog feels as if it's five human days long. And we could go on toying with this idea. But you see the point: The logic forced behind this idea is only one step away from the idea of subjective time. And subjective time undermines everything we thought we knew about time.

Here's some food for thought regarding subjective time: Why don't we measure the time we have left in our life?

This question brings us closer to our real feel and understanding of time. I remember when I was a child, and toward the end of the summer break, I would feel that nothing exciting happened anymore; the days just followed each other in an eventless, boring routine. Freedom felt like an eternity! I remember clearly what boredom felt like. But as I grew older, and I had more to do, time passed quicker and quicker. The older I get, the faster the days, weeks, months, and years go. Often I can't even say how I spent the past few years. Why did that year or month fly by so quickly?

As we grow older, we start pondering about how much time we've got left. And as we think about the future, we touch upon the concept of future and deeper issues of what we think we know about time itself.

So, what is the deal with the future?

Experiencing the speed of time going by, having a good idea of how much time is one hour or one year, why don't we panic thinking about the future? Why don't fifty-year-olds take the statistics at face value, and think, *Oh no, I only have twenty-five more years to live?* If we've already experienced the crazy speed of one year, why do we think that the remaining twenty-five times as much as this one year would go by slower? The twenty-five times one year will pass twenty-five times faster than the previous year.

If we really did have a clear sense and understanding of time, we would start dreading the time of our death as young children. People wouldn't be smoking or doing other harmful things to their health, because they would keep track of events that they have yet to experience while they are here on Earth.

Let me use a simple illustration. I'm walking from Carnegie Hall to Manhattan Bridge. When I reach the destination, I can decide whether I'll walk back or take the subway instead. It's a simple enough decision because I know exactly how much energy the walk requires.

Our recollections of past memories and emotions, albeit very inaccurate, suggest that we do have some sense of the past. The time to be experienced in the future, however, exists only as a concept. We are unable to measure the time that is yet ahead of us and consequently prepare for it in any way. The 2011 movie *In Time* touches upon the same idea. It is a brilliant depiction of what people would go through had they lived in this constant countdown. It doesn't aim to get a serious message across but it reflects upon what our feelings would be like if we knew exactly how much time we'd got left.

Keep your perseverance. You'll need it because this inaccurate and unreliable system called 'time' will reappear in the following chapters. At this point we can all agree that the concept of time is entirely arbitrary. The only reason why we say that there are twenty-four hours in a day is that we decided this when setting up our own system. We hacked and hacked this system until we reached the numbers twelve and twenty-four.[5]

We have the right to ask: How is it possible to live in this ridiculous system of time? How can we plan and schedule? How come there are no nuclear disasters every minute? How come satellites don't fall on our heads right after having been launched? We just can't plan accurately. If the rotation of the earth got slower, gravity would change. And with gravity, all processes would change, including length measurements, everything that is based on comparison. The direction of light would also change, making it impossible to measure time the way we used to. Our GPS navigation would get the distances wrong and splitting atoms in nuclear power plants couldn't be cooled in time, since measuring appliances couldn't use the same ground for comparison as before. How could we solve this?

The answer may sound paradoxical. Remember the story of the sundial and our inventor? We are doing exactly what he used to do.

---

[5] It was the ancient Sumerians who came up with the sexagesimal numeric system and divided a circle into twelve equal parts which made sense to them.

Instead of making the sun the point of reference, we use atoms, which seem to be working the same way hour after hour, and day after day—for now, anyway. We turn this into a general rule and assume that atoms function the same way in our solar system and beyond as they do here on Earth. We conclude that they are accurate and reliable. And because we said so, we expect them to comply. When we find that they are not accurate any more, we'll start over with a new means to measure time.

We've seen that time doesn't exist. It was created by our own inaccurate rules and is nothing but observing the expectations between two movements. And observation brings different results if we change the subject of observation in spatial context. The route of the events becomes shorter or longer.

The notion of time exists simply because we said so. That doesn't mean that anything we say about time would actually have an effect on it. No matter if we set rules, it will not budge.

1.2 Space-time

The previous chapter has shown us that the illusion we call time works accurately only where we measure it. The flow of time and the measurement itself must be in the same context, being exposed to the same conditions. This chapter aims to take it one step further and point out how the illusion of time is used to perceive space.

We'll see that time-tracking devices lose their validity if the flow of time and the measuring device are too far from each other, or they are exposed to different conditions.

Contemporary physics is slowly letting go of the conventional notion of time. Theories that once were taken at face value are said to be slightly untrue, although once they were the foundations of important statements that were awarded doctoral degrees and were generally considered true and proven. It seems we operate in a way where at first we state something, then we say it is true, then we look and search until we find a way to prove it. And when we find an error in our deductions, we suddenly declare that the theory previously considered as true is in fact false.

But how many tracks lead to the middle of the forest?

True can become false in a blink of an eye—science often contradicts itself. Here's a crude example of how our way of thinking has changed over time: If a thief was caught red-handed several hundreds of years ago, the rightful thing to do was to cut his hand off. Today, this would be considered a crime, and the one who cuts the thief's hand off would go to prison. Both actions are both right and wrong—it depends on the *then and now*. The only thing that has changed is the context: where and when, and according to what conventions we want to judge the context of the theft. The deed is the same, but the explanation is different; it changed according to the set of values of society, of people.

Truth is relative. Take the example of traveling from Nashville to Knoxville, Tennessee. Somewhere in the middle of a circa three hours' journey, there's a moment when it's exactly 10 am. But the next moment that's not true anymore; it's suddenly 11 am because I am in a different time zone. I arrived in the EDT time zone, having left CDT. And if I'm stubborn and won't stop saying that it's 10 am, I might even be called a liar, because I'm stating something that is not true. But, you see, everyone can be right: In Knoxville, I can say it's 11 am — according to the local time zone, EDT — and I can also say it's 10 am — according to the time zone of Nashville where I come from, CDT; both statements will be true.

Let me give you a third example. Robert H. Goddard published an article in the *New York Times* on January 12th, 1920.[6] In the article he presented the rocket theory, which would make it possible to leave the earth's atmosphere and travel in space. In 1920, Robert H. Goddard was wrong. 1920's physics said that thrust (T) can only be generated in the atmosphere. By definition, thrust is the force that makes an object move against air resistance; for example, the force of the air coming through the hole of a pierced balloon sends the balloon zigzagging across the room. Well, the *New York Times*

------

6 https://timesmachine.nytimes.com/timesmachine/1920/01/13/102738081.html ?pageNumber=12.

thought it impossible to generate thrust in space, since it is a void without air, and alas, there is no air resistance to overcome. Robert H. Goddard had to put up with mocking and rejection for a decade, and even some of the great thinkers dismissed his theory on the basis of fundamental physics. Goddard, however, went against the then-current truth and did not withdraw his theory. He died on August 10th, 1945, without receiving recognition for his work. Only after the successful launching of the Apollo 11 in 1969, did the *New York Times* admit their mistake and rectified their point of view concerning Robert H. Goddard's theory.

Everything is relative; there is no permanent truth or falsehood. The truth value depends on when and where we interpret a statement or an interaction—it is entirely exposed to space-time. Truth can easily lead to paradoxes; therefore we must handle truth with extreme care.

Here's the paradox of truth: *I'm not comfortable in an environment where I'm often*—let alone, always—*right.* I found that the more often I am proved wrong in a particular context, the more I will know about what I believed was true and others found false. This way, I can actually get one step closer to the real truth. My colleagues know me to be always saying, *"How I hate to be right!"* And I do feel that way. Because whenever I'm right, *ab ovo* I'm dismissing every solution and possibility suggested by others—although they might be more to the point than me. And therefore, I dismiss the possibility of the goal being reached in the best possible way. This goes against my own aim to find the best solution. Moreover, if I'm ignorant of others' truths, I won't have the chance to strengthen my own; therefore, I can't agree or disagree with others'. And also, by getting to understand other people's truths, I get to understand reality itself, as reality is made up of multiple truths.

The problem with the paradox is that I would have to know all truths of all possible external systems, and therefore, at that point, I'll be always right, cutting myself off any new knowledge, a.k.a. development. Therefore, I'll never experience the paradox, because getting to know all these many truths and statements would take considerable time, and thus, I'll have to put up with living in a world of half-truths.

But we are analyzing space-time, so it's time to bring up *speed*. Speed is defined as *the distance something or someone travels in a given time frame*. The everyday format of speed is mile per hour or kilometer per hour. But speed is a tricky thing. It has been observed that the faster we go a given distance, the shorter the distance becomes. Let me use an oversimplified and overexaggerated example to illustrate my point. Let's again travel from Nashville to Knoxville. We go by car at 90 m/h, meaning that we take the 180-mile distance in two hours. But if my speed is 900 m/h instead, then I travel only 170 miles, and I reach my destination in twelve minutes. Wait, what happened to those ten miles? This is confusing. Einstein himself must have felt confused. Distance can't decrease just because we travel really fast! And yet it does. Of course, not when traveling in Tennessee, and especially not by car. Scientists came up with a solution: If we must leave distance untouched, we'll modify the notion of time instead. The numbers must match always; 1 plus 1 must equal 2. But there are other concerning issues about speed and duration. If we take the previous example, those 170 miles are only perceived by the passengers of the car. For onlookers, it's still 180. If we agree with Einstein and say that distance does not change with speed, then it means that time passes differently to those who are sitting in the car and those who aren't. The duration won't be equal. Even if their watches show the exact same time at the departure, when they reach their destination the watches will show two different times.

We also use the expression *fabric of space-time* when we talk about space-time, which gives some sort of explanation as to why distance shrinks with increased speed. At the same time, it somewhat dismisses the theory that increased speed changes time, because when traveling the fabric of space-time, distance changes. We seem to have found a few contradictions. Let's go find a solution.

It is generally considered that the higher the speed of an object is, the more it warps space.

But how should we picture this space warp? Theoretical physicists and astronomists usually suggest imagining a sheet of paper that has two holes in it at one third of the paper's length. Should the paper lay on the table, and I put my pen above a hole, then trace a line to the other hole, I'll take a distance of about 4 inches; therefore, they are 4 inches away from each other. Then I start folding the sheet of paper and the holes get closer and closer to each other. The distance between the holes gets shorter. When I have half-folded the paper, its end pointing skywards, my pen would connect the two holes with a 2.8-inch distance. Then it becomes shorter and shorter, up until I've completely folded the sheet of paper and the holes seem like one. I've created a wormhole (i.e., a speculative structure linking disparate points in space-time[7]).

With the folding of the paper, the holes get closer to each other, the distance between them gets shorter, and so does the length of the route, and therefore, the time needed to travel that distance. This means that the time of getting from one hole to the other was still the result of the interdependence of speed and distance. (If you read carefully, you will find another idea about the space-time warp in the chapter on infinity, referring to it as an alternative or an inherent part of infinity itself.)

---

[7] Wikipedia article on wormholes.

The quite confusing thing about space-time is that time, as a notion of its own, appears yet again—and what is more, it is connected to movement. But in the previous chapter, we did a nice and thorough job of analyzing why time is unsuitable for use in environment-dependent calculations: It is relative. And there's another problem: If time behaves differently when the speed is higher as opposed to when it's lower, then how does time actually work? We can't specify it using self-defining theories or formulas, or ones that are built on wrong foundations. I think that none of the definitions where time appears as we consider it today can be accurate or true, let them be statements, calculations, or proofs.

At this point I would like to ask you to think about how you would define *space*. Not outer space—but the space around you. What does this word mean? What do we use it for?

I don't know of a single physicist or researcher who was able to point at space or explain why we even need it, let alone explain its elements, how it's made, and what the elements—quanta—are made up of. And it would be great to know the purpose of space, too.

One of the definitions of space goes like this: "Space is the boundless three-dimensional extent in which objects and events have relative position and direction."[8] Notice how this talks about what can be *done* in space, and not about what space actually is.

We might already have an answer at hand: Space is a product of coincidence; it is made up of coincidental parts. It doesn't have a purpose, since coincidence doesn't have one either. Therefore we live in an aimless space, and because we ourselves are the products of coincidence, we must be without purpose, too.

Is this how we feel?

In some scientific definitions, space is a mathematical model or a structure where objects relate to each other, while other definitions consider it a *void*. The definitions divert the question to a different field or simply don't relate to reality, because a new defintion could harm all those theories that are built upon it. We don't know. We have no idea what space or space-time is. We don't *see* it, yet we calculate with it and build upon it.

There are lots of theories about space and time, but there is no mutual agreement as to their basic definitions, and I believe we are thousands of years away from actually describing them as they really are.

For want of valid definitions, we have no other option but to work with Einstein's theories, those that we grew up with, those that

---

[8] Wikipedia article on space (physics)

people usually find suitable. This is not all bad, because we will be able to ask serious questions when we look at the foundations, the basic elements of our world, and the answers are bound to spark further questions, which might just lead us one step closer to reality.

1.3 Light

When we measure light or the speed of light, we do not actually measure speed, but instead, interactions: the effect light has on our measuring devices.

We say that the speed of light is constant, independent of the movement of the system within which we measure it.

If we shoot an arrow from a moving vehicle and it hits a tree, its speed equals the speed of the vehicle plus the speed of the arrow when it hits the tree. But if we use a ray of light instead of the arrow, then that ray of light travels with the same speed—roughly 186,000 miles/second—regardless of whether we shoot it from a moving or parked car. The speed of the vehicle does not add to the speed of the light since light is not part of our classical four dimensions.

We seem to find ourselves in a tricky situation once more: In order to be able to define light, we need a definition of time and space, or at least some sort of understanding. We've already seen that the interpretation of what time and space are, can only be made if we take into account how relative they are. Time proved to be like that—and we need it to calculate speed—and we weren't even able to put our finger on what space actually is. So we have two unreliable factors to define and describe a third one: light.

We might reach a seemingly correct result by following two seemingly correct methods. But would that be true? If we take literally that "the speed of light does not depend on the moving of the measuring system," then there's only one logical conclusion: The system in which light travels doesn't move. It is motionless.

A ray of light shot from a high-speed car and traveling in the same direction will have a speed of 186,000 miles per second. It doesn't matter where we measure: in the vehicle that goes at 100,000 miles per hour or before we start the car, when the vehicle is not moving yet at all. It's fascinating that no matter where we measure, in a speeding or a parked car, the speed of light will be 186,000 miles per second. We just can't generate conditions that allow one ray of light to overtake another if they had been launched at the same place, at the same time, even if something in their environment was also in motion. This leaves us with only two options: Either light travels in a static medium that does not move—or movement does not exist.

The medium in which light travels can be static in the absence of time, since we can't explain speed—i.e., motion—without the notion of time. And there's the issue of space as well. As we have seen in the previous chapter, it's impossible to put our finger on what space actually is, what it is made of, what it does, and how it functions. There's only one thing we know about space: It can interact with the objects that are in it, or it can have a one-way effect on them. Think, for instance, about what happens when we inflate a balloon: that particular bit of space gets divided.

Sometimes we can find the notion of time in space, for example, when we walk in a park and we seem to feel the passing of time. In other cases, time seems to be absent; just as we have seen with the speed of light. We say that everything spreads in space, including light, and still, light doesn't seem to tally with time; it doesn't care if it exists or not—because time doesn't affect the speed of light.

By the end of this chapter, we will reach a deeper understanding of light and its true nature. To this end, let's take everything at face value for the time being.

You might be asking: how is it possible that in some cases time exists and in others it doesn't? Fair question. According to what we've found this far, we must create a model in which both statements are true. Later on, in the chapter on dimensions, I will propose a model where everyone will be right, in which there will be time—and in which there will be no time.

Both time and space-time were found to be our own inventions. We also found them relative, and experience led us to inaccurate definitions. But what about light?

If only we could say that we know everything there is to know about light! It is one of life's basic components after all. Unfortunately, we must admit that light is yet another of our arbitrary constructions. We endowed it with features based upon our perceptions. It is colorful or white, warm or cold; it drives the darkness away. Saying that light can destroy and it can heal, I seem to contradict myself, but both statements are true. And if we still hold on to the notion of light as we know it, we can also say that there is light that is visible to the naked eye, and light that isn't.

What we find true today is that light is made of photons. But there were times when we were sure that it was made of corpuscles (small particles) —in 1660—and these are of different colors. That's because even then we saw that white light splits into seven basic colors. And if you find this funny, let me assure you that in four hundred years we'll find the theory of photons equally funny. We don't know much more about light than we did back in the 1600s. Sad, but true.

No one ever saw a photon.

Just as no one ever saw a corpuscle.

It was because of observed interactions that led us to believe that photons exist, the same way we believed that corpuscles exist.

Light is divided into visible and non-visible light, only because there is a part that does not interact with our relevant organ—our eyes—or with any of those devices that we use to measure interactions with. But is it true that it doesn't exist just because we don't see it?

Well, yes and no. We're getting close to the issue of relativity once again… and its conclusions regarding time and space-time were not pretty. Still, we have to embrace relativity, because certain insects and mammals have the ability to perceive a part of the light spectrum that is invisible to us humans. And the opposite is true as well: There is a part of the light spectrum that we see and other creatures don't. What kind of image is there in their brain upon seeing a blooming almond tree? Insects might feel sorry for us because we can't see those brilliant ultraviolet colors. We feel sorry for our cat or dog because they miss out on so many of the magnificent colors. Yet all insects, cats, dogs, humans, and all animals live happily in the beautiful light they are used to, that they think is the reality.

Let's see how science defines light: "Electromagnetic radiation (EM radiation or EMR) refers to the waves (or their quanta, photons) of the electromagnetic field, propagating (radiating) through space, carrying electromagnetic radiant energy."[9]

Light is considered to have no mass in stationary state. The problem is that we have never seen or measured photons in stationary state. The solution might be found in acceleration, since the faster we want to go, the more energy we need for acceleration. Today, science says that energy is mass, according to Einstein's famous theory: $E = mxc^2$.

The following illustration summarizes the issue with acceleration and energy—that is, how much fuel is required to reach a certain speed. I set upon a 60-mile journey by car. At a speed of 60 miles per hour, my fuel consumption is 1,7 gallons per mile. If I travel the same distance at 72 miles per hour, my car consumes 2,1 gallons per mile instead. Not bad! If I go 12 miles per hour faster, it only takes 0,4 more gallons of fuel per mile. So I accelerate. Now I travel the same distance at 96 miles per hour and expect to use 2,9 gallons of fuel per mile. But lo and behold, my car has consumed 3,8 gallons! What happened? Well, interestingly enough, at this speed, such acceleration requires not 0,4 but 0,75 gallons of fuel.

Let's say that my car has superpowers and the highway is also of superior quality, so I can travel the exact same distance at a speed of 200 miles per hour. To my utter amazement, at reaching my desti-

---

[9] Wikipedia article about electromagnetic radiation.

nation, I find that my car has consumed 15 gallons of fuel. Remember, a 12 miles per hour acceleration speed required 0,75 gallons of fuel at 96 miles per hour driving speed—now it is around 1 gallon. Let's not stop just yet. I go as fast as I can with my supercar and reach 720 miles per hour. I wonder how much fuel my car consumed. At the end of the 100 miles journey, my car has consumed 1700 gallons at this mind-blowing speed.

Try to visualize it: 1700 gallons of fuel weigh approximately 8800 pounds. My car has to carry this weight to be able to reach 720 miles per hour. And the more weight it carries, the more fuel it consumes.

We quickly reach the physical edge of possibility; this self-perpetuating cycle stretches the limits. The faster we want to go, the more fuel we need—disproportionately more. The graphic representation of the relationship between speed and energy consumption would be a degressive diagram.

But let's not stop driving. Putting all scientific possibilities aside, let's make our car go 10,000 miles per hour. How much fuel would it consume? Well, at this speed and with this amount of weight to be carried, our car would consume the entire fuel resource of the planet in a matter of moments.

There's no need for further explanation: It's evident that if light had mass, it would never reach the speed of light. It would require more energy than is present in the universe.

However, there's a slight contradiction in that light can have an effect on objects that have mass. Experiments with a particular spacecraft called solar sails have shown that light affects objects and it can even increase their speed. Even if we had inexhaustible energy resources—but we obviously don't—the speed of light would still be a puzzle. Such fast traveling would require an engine that generates the speed of light *and* a tiny bit of thrust or pulling power for an object. However in our dimension, based on our present knowledge, nothing can surpass the speed of light, so this task has a theoretical obstacle.

So how does light work? I loved to play with colorful marbles when I was a child. The rule of the game was to hit a marble on the ground by rolling another marble from a 1.5-yard distance. The second marble rolled toward the first one, hit it, and transferred its swing. The first one started rolling and the second one stopped. We expect something similar to happen to light too. There's a difference, however: Marbles have weight and mass, while light has neither.

And yet it has an effect on mass.

But light is energy! It's as if I've been trying to compare an apple to a mouse pad all this time… If light is, in fact, energy, then that explains why it affects mass: Energy is converted into mass.

The definition of energy goes as follows: "In physics, energy is the quantitative property that must be transferred to an object in order to perform work on, or to heat, the object. Energy is a conserved quantity; the law of conservation of energy states that energy can be converted in form, but not created or destroyed."[10] Although the definition talks about *where* to look for energy, it does not clarify *how* energy actually works on an elementary level. Imagine witnessing energy becoming matter, how exciting that would be—it doesn't even have to be creating material, i.e., something that creates energy, like a burning piece of wood. How exciting to see the very moment of transformation when energy turns into material and we can actually take it in our hands.

We haven't turned energy into matter yet, even though the definition allows for it to happen. Light has wavelength—indicating the distance between waves—and frequency—representing the periodicity of waves. It is wavelength that defines the type of light or electromagnetic radiation. For instance, what we perceive as colors are, in fact, different wavelengths of a particular spectrum of 380 to 780 nanometers per second. The wavelength of the color red, for

---

[10] Wikipedia article on energy.

example, is 620 nanometers, and it has a 4.8×10^{14} Hz frequency. For the sake of comparison, the frequency of a deep note on a drum machine is 70–80 Hz (there are no thirteen zeros behind this number). That's an astounding frequency for one second!

We've quickly arrived at the issue of wavelengths and frequencies.

Even though we've never seen an oscillating photon, we believe that it's their wavelength that determines whether that particular light is visible to us or not. What makes anything visible? We have already mentioned that the sight we see must be able to interact with our optic nerve. Imagine a wire mesh of one square yard fastened onto an empty box. In this particular example, the wire mesh represents the human eye's retina. The rows of the mesh are three inches wide. Let's hold a handful of sand over the box, where the sand represents ultraviolet light. We scatter the sand through the wire into the box. The grains of sand slip through the mesh without any difficulty, and all of the sand lands in the box. A few grains might have touched the wire before getting into the box, but most of them didn't, since they were too small to be caught by the mesh. So we can conclude that there was no interaction—or if there was, it's insignificant. There's no way of telling how many grains of sand collided with the mesh. So, considering that the mesh was our retina, we didn't see the ultraviolet light.

Now let's throw a brick onto the wire mesh. It will bounce back. What an interaction! It illustrates how our optic nerve perceives red light: There is interaction and we see the color red. Let's not forget that light has no mass, so nothing bounced; our eyes perceived electromagnetic radiation or energy turned into mass, and so, we've literally had a photon slam into our optic nerve.

Today's science explains the process by photons moving in a wave-like fashion while traveling at a constant speed. It's like stretching a rubber band and then tugging its end. The distance between waves is the wavelength, while frequency refers to the rate at which they occur—i.e., how often we tug the rubber band within a given time unit. Perception is made possible if the wavelength is long or short enough to interact with the corresponding sense organ. The mesh example has shown us how interaction works: If particles are too small, they fall through the wire mesh, but if they are much larg-er—say, the size of a house—they will not even come into contact with the wire mesh, and there will be no interaction.[11] We need particles of the right size to enable interaction with the wire for sure. The visible wavelength in our example was the brick.

The previous paragraph contains another key statement: We meas-ure everything by analyzing interactions. What we call measure-ment is but an endless cycle of comparison. Even so, those wave-

---

[11] Wikipedia article on visual perception.

lengths and frequencies that are not able to interact with the materials and energies known to us are nonexistent to us. We know nothing about them; therefore, we don't know how to measure them either. William Shakespeare pointed out the truth in *Hamlet*: "There are more things in Heaven and Earth, Horatio, than are dreamed of in your philosophy."

We know that waves interact with each other in two different ways. Think of sea waves; crests and troughs follow each other, sometimes in degrees, sometimes in sequence, one after the other. When two crests or two troughs meet—i.e., they are in phase—they strengthen each other, so to speak. In this case, we can speak of constructive interference.

When waves are out of phase, a crest meets a trough, and the opposite happens: They weaken and can even destroy each other. This is what we call a destructive interference; the wave collapses.

Waves behave like this not only at sea but in every other case when a phenomenon is wavelike. Therefore light works like this, too.

Light as energy is to be further discovered, though, because it has properties that we still don't understand, up to this day. Light, as we know it, is like a wave: It has a wavelength and a frequency.

If we take a piece of black cardboard with a hole in the middle, face it toward the wall, and direct a lamp toward it, we will see the clear shape of the hole projected on the wall.

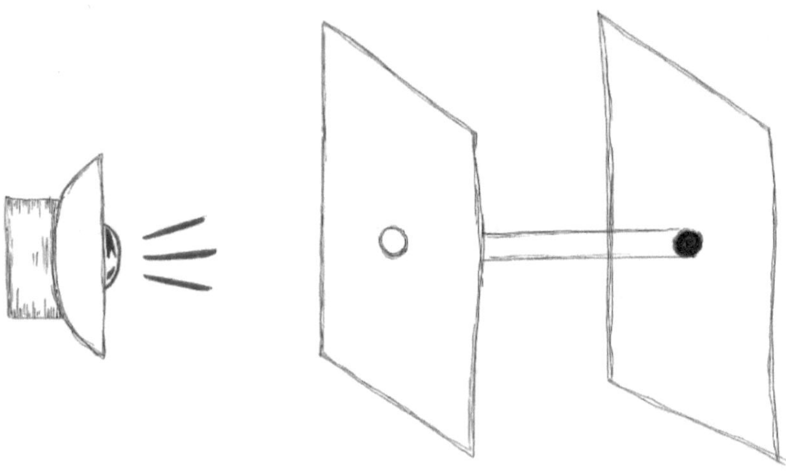

But sometimes light doesn't behave as we expect. There's the well-known double-slit experiment to show it, which still questions our knowledge of light. This is how it goes. Take a piece of cardboard and cut a 0.3 x 1 inch rectangle into it. Now take the graphite refill of a drafting pencil and place it in the middle of the hole perpendicularly to the rectangle's length. Direct laser light—a laser pointer—to the hole, or rather on the graphite in the middle and it will create the following projection on the wall:

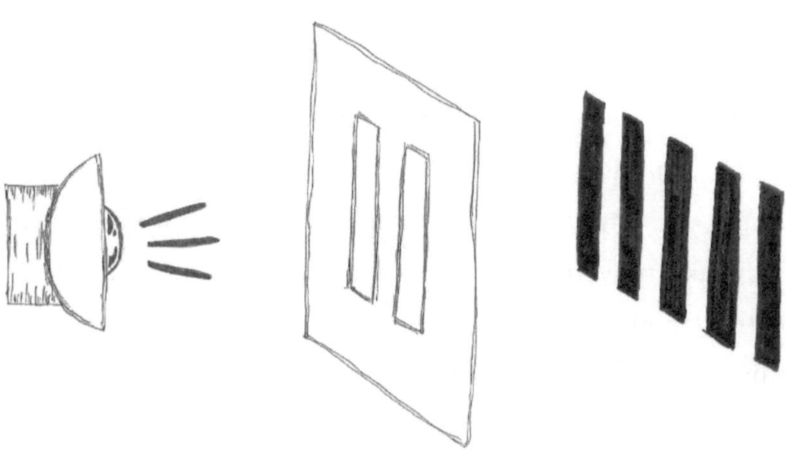

It is slightly confusing that we see more stripes on the wall; a single thing should have only *one* shadow. But we can explain it with wave phases. All points of the wall or screen receive light waves from both slits, but the waves of light emitted by the laser pointer travel different distances, so they weren't all in the same phase. Because of the interference, some waves strengthen, others weaken each other. That's why there are lit and lightless stripes.

The double-slit experiment was the first to prove that light is a wavelike phenomenon. But light has a dual nature: It can move both like a particle and a wave. As technology developed, the experiment was repeated with a small twist. Only one photon at a time was let through the slit. Once it reached the wall, the next photon was sent. This was repeated several times, and the researchers waited for a picture to be drawn.

The projection looked exactly like the previous one, when a single ray of light (that is, all the photons) was sent through the slits. But how could this happen? True enough, when several photons travel in a wavelike movement, they overlap, and through interference, they either weaken or strengthen each other. But this time there was only one photon. How could it interfere with itself?

We'll digress a little so that we can understand the results of the double-slit experiment. We'll step into the realm of quanta.

Quantum is the smallest amount of a physical quantity that can exist independently. In physics, this translates into the smallest amount that can be added to a measured amount, where we can measure this addition. For example, photon is the quantum of light, but this doesn't mean that it is its most fundamental element: A photon is merely the smallest unitthat we can measure and thus find if it may affect the result of a particular measurement.

The result of the double-slit experiment with one photon can be explained by combining quantum physics and the theory of relativity. If we look at the inertial frame of reference of light, we find it lacks one dimension: time is absent. From the light's perspective, there's no past and present, only the *now*. We people need speed to travel from point A to point B in space. And speed inevitably entails the presence of time. If we take out time from the formula of speed, what we're left with is distance. And we can't travel from point A and point B, because we don't have speed; we are stationary. There is, however, one exception where we don't need time: We can send information from point A to point B in zero time.[12] Quantum entanglement is by now a fact proven by practical experiments.

Unfortunately, we have no true, eternal knowledge of the operation of quantum entanglement. We know some things about *what* it entails, although the *way* it works is still a mystery. Still, that is enough to help us reach a deeper understanding—or non-understanding— and to get closer to a world that contradicts whatever we've been taught.

Of course it's an exaggeration, but to make my point I can say that photon splitting is an experiment we can try at home, and it entails quantum entanglement. Picture a single particle of photon embodied as a ball of light. This light ball goes through a nonlinear optic crystal that has a special refractive quality, resulting in two light balls. They are almost identical, the only difference between them being that one is spinning clockwise and the other counterclockwise. Let's imagine we put the two balls next to each other, two feet

---

[12] Gergely Nagy and his team have been working for years on a mind-blowing study, entitled *Kvantumradar* (Quantum Radar). If you are eager to find out more about the realm of quanta, I heartily recommend diving into this comprehensive research.

apart. Let's stop the ball that's spinning clockwise and spin it the other way round. Once we change the direction of a ball, the other changes too, even though we never touched this second one.

Now, let's pick up the light ball spinning clockwise and throw it at an imaginary piece of black cardboard approximately four feet away. The light ball disappears as the black cardboard absorbs it. And once this happens, the ball that's been spinning counter-clockwise vanishes too, as if that has been absorbed as well. It simply disappears. And this is what we call quantum entanglement.

Two entangled particles experience exactly the same events in every case. In fact, we shouldn't even talk about it as two separate particles, but one with a *nature of two.* Quite a paradox, just like Theseus's ship: If all its components have been replaced at least once, is it still the same ship?

Let's jump back to our two light balls for another moment. What would happen if instead of the black cardboard, we threw one of the balls at a piece of cardboard that happens to be on the Moon? Well—the same. As the ball spinning clockwise smashes into the piece of cardboard on the Moon, the other ball, spinning counter-clockwise, which was right before our eyes, disappears too. The point of this example is to demonstrate that no matter which cardboard I throw the light ball at, the other ball in front of me disappears with zero time sequence.

Quantum entanglement is a practical example to show that time, as we know it, doesn't exist everywhere. In our understanding, without time there is no space; we wouldn't be able to travel in it because speed is necessary to change position, and speed can only be defined with time. If there's space but no time, we only have curvature of space: Traveling, in this case, does not require time at all.

It is possible that time exists at quantum level, although it works in a completely different way than time as we know it. But one thing is clear: When the direction of the spinning light ball changes—whether on the Moon or at the edge of the universe—the direction of the other light ball's spinning here on Earth changes too, at the same time, i.e., with zero time lapse. It doesn't matter how many billions of light-years the two balls are apart from each other. If we look back upon the double-slit experiment, the theory may be acceptable that the photon directed at one of the slits did cross both slits at the same time. Photons are able to do so, since their dimension doesn't contain time, and the question of whether two things happen at the same time is unfounded if time doesn't exist.

Light, as we perceive it day by day, is something extraordinary. I'd like to refer back t few chapters ago where I said that the more complex a structure, the more simple its appearance is. Light fits the bill exceptionally well.

As we step into the realm of quantum, the world that we know turns upside down. There's no past or future here. We only have *now*: Everything is happening *now—everything at the same, nonexistent time.*

1.4 Infinity

The notion of infinity, like other definitions before, requires us to disregard our perception of reality. Let's think differently than others do! The idea of infinity will be the first crossroad on our journey, where we can decide between options on how life may have started and matter came into existence. Following logic, we'll only have two options—therefore, the choice may not be that hard.

To get to know infinity, let's start with the question: *Where* do we find it?

The question above is unanswerable, so let's look at it from a different angle. We should first accept the following statement:

If we can't find a single place where infinity is absent, then we can declare that infinity exists everywhere. If we do find such a place, we can draw a perfectly accurate outline of the points where infinity was absent. In the end, we will have a map of places where infinity exists.

If it seems a bit confusing, we should read it again, and the logic will be seen—and this logic will help us delimit the locations of infinity and finiteness.

Regarding light, we have already seen that quantum-level worlds have rules of their own; such as the absence of space-time as we know it. So, if we want to analyze the notion of infinity in the context of space—because we believe that infinity doesn't exist without time—then it only makes sense to analyze infinity in our own dimension. As we IT developers say when an error occurs in the control of a certain program, "There is no infinite cycle, but finite time for execution."

If we imagine that we step out on the street, our mind immediately transports us to the first crossroad. We can imagine ourselves landing on the Moon, or we can fly to the star closest to us, Alpha Centauri C—although it's not even visible to the naked eye; it is 4.23 light-years away from us. (One light-year means the distance light travels in a vacuum in one year at about 186,000 miles/second speed. This means that it travels around 5,879,000 million miles per year.) We need to imagine that it takes 27,867,000 million miles to travel to Alpha Centauri C. If we traveled by car, driving day and night at a constant speed of a whopping eighty miles per hour, it would take us more than thirty-five million years to reach our destination.

It's also possible to take a trip to the neighboring stellar system by using our imagination. The Andromeda Galaxy is only 2.5 million light-years away. Keeping the speed of eighty miles per hour with our car, we'd be there in just 20, 769,230 million years. There are around 200,000 stars in the Andromeda Galaxy. It looks like one dim star from here below, but in fact, the ray of the Andromeda is 110,000 light-years long—i.e. it takes 110,000 years for light to travel all the way until it reaches us.

If we travel in outer space, distances will simply blow our mind, so we might as well forget about converting them to our miles or kilometers. Even a trip to the neighborhood feels endless—because of the enormity of the numbers.

But let's go on. Let's cross those 100 . . . 800 billion galaxies, each probably inhabited by a couple of hundred billion stars, until we reach the edge of the observable universe, which is just about 14 billion light-years away. Of course, I can't offer you the absolute truth about stars and galaxies. Recent discoveries estimate the width of the universe is around 78 billion light-years—which shows the relativity of truth again. The size of our universe depends entirely on the time of measurement; we always consider the most recent statement to be true.

But the reality is, we have no idea what the size of the universe is. So we'll just travel to its end in our imagination. Because numbers are so high here that they resist rationality, there is no point in keeping count of the billions of light-years. They will not help us understand the immense distances. The point is that as soon as we leave our universe behind, we're bound to find another similar or dissimilar universe. And when we cross that universe, another will follow, and then another. The idea of multiple universes (i.e. multiverse) comes as close to understanding infinity as it can get. New universes will keep coming; this makes infinite space real—although this is again something we have hardly any idea of, as we pointed out in the chapter on space-time.

We've had a sublime trip, but we're still far from understanding infinite space. Let's look at it from the opposite perspective, and see how infinity decreases.

Picture a toothpick that we pick up in our hand. Now let's break it in half. We'll have two half toothpicks. We take one of these, and break it in two as well; we now have a quarter of a toothpick. Let's set the limit of repeats to infinity. As we keep breaking the pieces in half, we have 1/8, 1/16, 1/32, 1/64 pieces of toothpick—and it goes on until we reach an infinite number of pieces. In mathematics, it's all nice and easy, but considering the toothpick has actual material it's made from, we near the impossible task. After having broken the original toothpick into so many halves, we'll get to a point when we are left with a single atom.

But we don't stop. Let's cross the electron cloud of the atom and find the nucleus. We split it in half. We can do it and we find one proton and one electron. We take the proton and cut it in two, getting two half hadrons; one contains a quark, the other a gluon. Gluon is some sort of glue, which holds the quarks together, so let's slice the quark in half instead and see what we find.

Quantum physics defines quarks as elementary particles. I use the expression *cannot be reduced any further* when I think of them. But if it's not made up of further things, how did it come to exist? I want to find out, so I will take this particular quark apart, whatever it takes.

There are only two possibilities:

- I will either find that the quark contains one or more other particles, or
- I will find literally *nothing* in them.

If I do find other particles in the quark, that means that quarks are not elementary particles, after all, so I can go on with my seemingly endless experiment and split whatever I find in half—mathematically, that's completely possible. Nevertheless, I will inevitably reach a particle that contains nothing, sooner or later.

But what if I find nothing in the quark, in the elementary particle? It only seems logical to ask, did my toothpick, therefore, come from nothing? I have only two possible answers again.

- Option A: If something comes from nothing, it means that it was created. And creation involves an omnipotent higher power.

- Option B: If I keep finding further particles that I believe to be elementary and go on splitting them in half, I will gradually descend into deeper and deeper levels of infinity. This means that the toothpick is made of an infinite number of small parts, and therefore it is impossible for it to actually come into existence.

We have no option C in this case. If physics confidently claims that there is such a thing as an elementary particle, in my opinion, it actually reinforces the notion of *creation*. An elementary particle must be made of something else—or else we are dealing with a particle that can only be reduced to itself, or to be more specific: It can only be reduced to nothing. This makes it unsplittable, but I still need to prove how something can come from nothing.

It is possible to analyze infinity from the perspective of light. By now we know that light works as if it belonged in a completely different dimension, but it's hard to understand because we perceive it here in our dimension. Let's try to find an explanation!

I split a photon—the elementary particle of light—into two, creating quantum entanglement. Then I launch one half into outer space. It will immediately be absorbed, with zero time required, because darkness absorbs light. Wait a minute! Didn't we say that the speed of light is 186 000 miles per second? Indeed we did, but we measured it according to our own perception of time. But light doesn't work that way: A photon reaches its destination at exactly the moment that it was launched, regardless of the distance it has to travel. The key is quantum entanglement: Both halves of the photon function in the same way. The one that is close to us and the one that is far away have one common fate, no matter what. Because the two halves of the photon are entangled, there's only one option: The far-away photon half immediately—according to our perception—forwards the information to its other half residing on Earth. It would be simply impossible to have one half of the entangled photon orphan wandering around here for billions of years. So just as its counterpart reaches the edge of the universe and crashes into a planet, its sister here on Earth disappears as well.

Entanglement is a promising logical possibility. Imagine sending halves of entangled particles into all directions of the universe. Our task would only consist of observing the behavior of the half that remained on Earth. If it disappears, it means that its counterpart crashed into something. If it doesn't disappear, but we observe certain changes to it, its parameters get modified or it interferes, for instance, it suggests that its traveling brother has not been absorbed yet. This, of course, doesn't necessarily mean that it encountered infinity itself.

If we can detect the places in outer space that absorb the entangled particles, in due time we will also be able to map out the places where there is no absorption—and that might just be infinity itself.[13]

We don't know yet what happens to a half-particle residing on Earth if its entangled counterpart in outer space encounters infinity, but there are some promising results already. It will be another few years or decades before we get more accurate results.

It's hard to think how an entangled half-particle behaves in space in the absence of time as it approaches infinity. How can it travel in infinity—whose distance is infinite light-years? It would still take an infinite amount of years to travel that kind of distance, regardless of its speed, because it doesn't matter if it travels slowly or infinitely fast.

Mathematics did a nice cover-up of calculations that involved infinity. There was a time when upon using zero in a calculation,

---

[13] Gergely Nagy's quantum radar theory and experiment.

serious punishment was applied, or when they couldn't handle zero, they simply stopped using it; it didn't exist. Certain limits were introduced in mathematics so that anything that went beyond the limit was simply dismissed. For instance, if the result of a calculation was something similar to 1.5599...99, it was automatically supposed that the decimal digit would always be 9, going on for infinity. Regardless of whether this is correct— it might be— further calculations used 1.56 as the rounded result instead of the accurate number. However, calculating with billions of light-years can bring unpleasant surprises. Nevertheless, we've channeled this intuition and others into rules and formulas, creating limits to certain values, so that we wouldn't have to bother with infinity, where it is of no consequence. This way, the calculations are not entirely accurate but the numbers add up.

There's the definition of countable and uncountable infinity too. In 1870, mathematician Georg Cantor[14] came up with the so-called set theory, which differentiates between sets with infinite elements and sets with finite elements. Sets with infinite elements are further categorized into countably and uncountably infinite sets.

---

[14] 1845–1918.

Here's an example of a countably infinite set. Although it will not get us much closer to understanding infinity, at least it allows us to see how mathematics deals with it.

Numerator of the number	Previous number + 1	Previous number +2	Previous number +5
First number	1	1	5
Second number	2	3	10
Third number	3	5	15
Umpteenth number			
Infinity with the applied rule (The numerator of the number will show the exact ID number of the infinite element of the sequence.)	$\infty+1$	$\infty+2$	$\infty+5$

As we can see, it is possible to reach the last number of the sequence. And if we had the necessary time—that is, infinity and then some—to reach the end of the sequence, we would know for sure that we got the final number by adding 1, 2, or 5 to it. So we can predict the result of infinity—so, it is countably infinite.

Hilbert's paradox of the Grand Hotel takes it to the next level and shows how an uncountably infinite set works. It demonstrates that a fully occupied hotel with infinitely many rooms may still accommodate additional guests, even infinitely many of them, and this process may be repeated infinitely often, for example by moving all guests to the room whose number is twice as high as their original number. Uncountable infinity basically means that we can't describe the last number of a sequence; each element is built on the previous one. Let's start with the following table:

Numerator of the number	Infinity generator
1	0.2̲3456789
2	0.54̲859912
3	0.895̲42578
4	0.4687̲4263
∞	Decimal number with infinite combination, where the last fraction digit is underlined, the last one meaning the one in infinity

Let's think further. We have an infinite sequence above, based on which we can create an uncountably infinite set. We'll take each underlined number, and make a new decimal number out of them. It will look like this: 0.2457… Now here comes the tricky part of actually generating the uncountably infinite number. We'll add 1 to each decimal digit. The result is 0.3568… There's at least one digit in this decimal which is different from the numbers in the sequence from the chart. And if we add another row to it, we can also add one more fraction digit to our decimal number. Then our table would consist of $\infty+1$ rows. And even if we had all the time in the world—say, infinite time and then some—we would never be able to get to the final number of the sequence, because the new number would always be defined by itself.

In geometry, a circle is the notion closest to infinity, and I don't refer to the continuity of the outline but rather its formation. It's more complex than we'd initially think. The common definition of the geometric circle is: "A shape consisting of all points in a plane that are a given distance from a given point, the center."[15]

Let's take a look at the definition of the geometrical point, as it seems crucial to understanding the one before: "Geometric points do not have any length, area, volume, or any other dimensional attribute."[16] A point is basically a zero-dimensional location.

We find that a geometric point is of a size that doesn't exist, and yet we use it to define the place of a circle.

---

[15] Wikipedia article about the geometric circle.
[16] Wikipedia article about the geometric point.

What happens if we work with a notion that cannot be logically interpreted? If the whole definition is fiction, the result can't be real, right? It is not at all logical to get to a correct result using two faulty partial results.

Therefore, calculating the area of a circle will be fun. We'll use the number pi. Its value has been a mystery for centuries, because although technology today allows us to enumerate billions of decimal digits of the number pi with exact precision, even so, it is still not accurate. After all, we still haven't reached the end of the sequence. But why can't we define pi exactly? Perhaps because we don't actually know what a circle is made of.

Let's grab an imaginary compass and draw a circle, supposing that the very tip of the graphite is the size of an atom.

Once the tip of the pencil touches the paper, we see the atoms of the graphite lining up nicely, next to one another. They all have a beginning and an end.

Between two atoms there is space. We can trace a line from one atom to the other, and can also connect the atom with the center of the circle. By drawing these lines, we get a triangle, whose area we can calculate accurately. If we connect enough atoms along the edge of the circle, we'll get a polygon, which can be reduced to a finite number of triangles. We've checked the tip of the compass when we started drawing: There was no curvature, only a sequence of atoms, lined up next to each other. If there are a lot of them and they are squeezed tightly together, we perceive them as a circle.

I believe that once we get the definition right, it will lead to correct result, so let's try and find it. I'll calculate the area of a forty-unit-circumference circle using two different approaches.

First, we'll do a quick calculation with the formula for a circle using pi: $\pi \times r^2$. The result is 127,32395447351628.

The other approach is the polygon theory, depicted by the chart below. The result is almost the same, but it lets us see the correctness of the logic behind the calculation. The more triangles we draw within the circle, the more accurate the result gets.

A geometric circle is an illusion. And where there's illusion, there's relativity. But once we have a comprehensive definition and see how we can make it work, calculating a circle's area can be done accurately.

Let's take the example of a big circle in space, whose diameter is a couple of billion light-years. Even if we used the most recent and most accurate value of pi, our calculations would most certainly be wrong by a couple of billion square kilometers.

Number of triangles making up the circle	Length of each triangle's side along the edge of the circle	Area of the circle
4	10	100
8	5	120.710678
16	2.5	125.683487
32	1.25	126.9146298451107668
64	0.625	127.2216726561699289
128	0.3125	127.2983871002603138
256	0.15625	127.3175628227283909
512	0.078125	127.3223565728515325
1024	0.0390625	127.3235549991020861
2048	0.01953125	127.3238546049597346
4096	0.009765625	127.3239295063800682

8192	0.0048828125	127.3239482317324018
16384	0.00244140625	127.3239529130703147
32768	0.001220703125	127.3239540834048
65536	0.0006103515625	127.3239543759883929

If we calculate the same area by using the polygon approach, we only need to define the number of triangles that the polygon can be divided into. It's entirely possible to do that by measuring the length of the triangles' sides. And we could even measure that in space—well, if we had accurate knowledge of space, that is. We said that the circle is an illusion. If we use the method correctly, however, the notion of infinity disappears altogether.

The way distance is included in the notion of infinity is another intriguing issue. Distance is defined as the length of a straight-line segment that links two points.

You see it right: This definition uses the notion of *point*, too—the point that has no size, or to be more specific, its size is zero. So, literally speaking, distance is the space between nothing and nothing, and if I want to be really accurate, I have to start measuring precisely from the edge of that zero-sized object. Further than the logical pain of defining the point, and that the definition of distance didn't really exact anything, what we get is not even the reality of the definition. Even though we defined distance, we don't use this for measuring, so it can't be accurate.

Measuring distance is a problem as old as humankind. In the golden days, people measured distance by associating it with certain body parts, so they invented digit, cubit, foot, and the like. We've already talked about inaccurate units of measurement that are based on comparison. But talk about inaccuracy and comparison when it comes to body parts! It's only natural that one person's arm or foot has longer bones than the other's, so these units can't be of the same length. Some funny situations could have been witnessed on an ordinary day at a marketplace. When a 5 ft 10 tall man walked up to a 4 ft 8 tall merchant and asked for 10 cubits (combined length of the forearm and extended hand) of fabric, how much did he get?

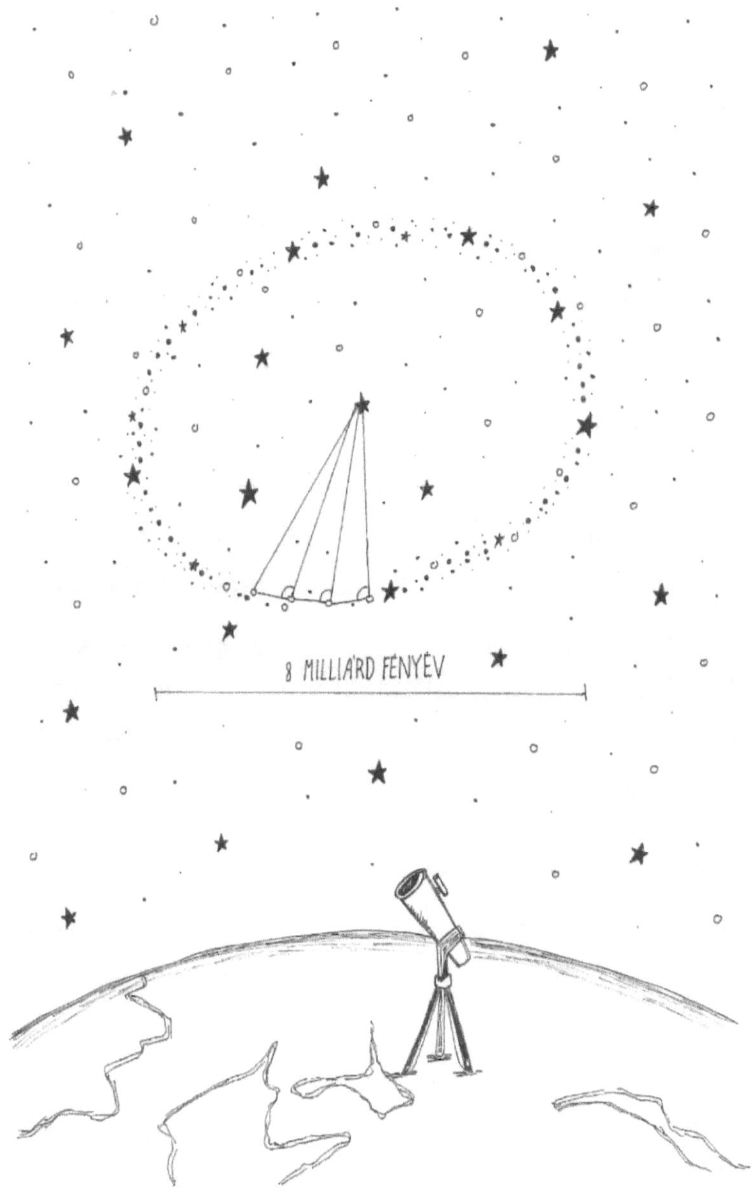
8 MILLIÁRD FÉNYÉV

Did the tall customer get as much as he wanted? And what if only tall customers bought fabric from the short merchant? Did he go bankrupt after a while? In any case, there was an urgent need for unity and constancy when it came to measurement units.

Notice how we build on an inaccurate unit of measurement, we force the use of it, even after we see it's not fit for its purpose.

Since actual elbow length (cubit) differs from one person to the other, initially they probably calculated an average. Two thousand years and longer ago, a cubit was roughly 1 ft 5.7 in. (Or 45 cms, to use the metric system—which is also food for thought, we'll come to dissect that as well later on.) A cubit had clearly nothing to do with the actual length of anyone's arm anymore, seeing that my arm is not as long as yours. A standard must have been made of either wood or metal, which was called cubit from then on and served for comparing length.

Around 300 A.D. a new problem arose. A merchant could have been in trouble when serving a customer who knew the new cubit. It was at this time that the standard length of cubit was changed from 45 cms to 55 cms (i.e., from 1 ft 5.7 in to 1 ft 9.6 in). We may find it funny how measurements can be associated with body parts, but it surely is over the top that from one day to the next, 10 cms were added to a measurement unit that people thus far considered true.

Come modern age, in the 1800s, one meter was defined—as the earth's circumference around the prime meridian divided by 40 million. But the earth's surface changes continuously, and consequently, so does the length of the prime meridian. They could've just taken a piece of wood and declared: *This* is 1 meter. It would've been all the same.

But they needed something to rely on, something that could be presented in a scientific-sounding way. The point is to do with relativity; something constant was defined by a variable (the changing Earth). So this unit of measurement was doomed from the beginning. At least they realized it early on, so they decided to make a standard: A so-called meter bar made of a special alloy, ninety percent platinum and ten percent iridium. The length of the bar proved to be constant at a given temperature. And that's the key: We can talk about that bar being one meter long only when the environmental variables have the same value as they were when the standard was created. In addition, no interaction is allowed, i.e., no change that could modify the percentage of the alloy and consequently modify its length.

The history of the standard meter bar is as unfortunate as it can get, the exact length of one meter has been compromised over and over. For instance, in 1948, the krypton standard was introduced as a law, which defined meter as the length equal to 1,650,763.73 times the wavelength of the radiation of the krypton-86 atom corresponding to the transition between the levels 2p10 and 5d5 in a vacuum.[17]

Let's realize how we can't let go of our own definition of length associated with the prime meridian—like a dog that grabs its bone. We prove using atomic wavelength that one meter *must* be one meter. Does this ring any bells? The synchronization of UTC and the earth's rotation speed might come into your mind from the chapter on time, perhaps.

Unfortunately, the krypton standard proved inaccurate and soon failed. This happens if a theory cannot meet the requirements— and in the case of the meter, that need was accuracy. In 1965, a new method was devised that would finally be worthy of our trust: Up to this day, we define the length of a meter by two other phenomena whose function we have no eternal true knowledge of.

One is light.

The other is time.

The new definition of meter is the length of the path traveled by light in a vacuum during a time interval of 1 / 299,792,458 seconds.[18]

---

[17] Wikipedia article about the history of the meter.
[18] Wikipedia article about the speed of light.

The definition was slightly modified for the sake of precision in 2011, at the 24th General Conference on Weights and Measures. Since then, a meter is to be defined by a physical constant. The revisited definition is the following: *Meter is the unit of length. Its sign is m. Its length is determined by the speed of light in vacuum as a physical constant which equals precisely 299,792,458 and whose unit is m/s.*

We can't help but notice that the resolution of the 24th General Conference on Weights and Measures defines light as a physical constant.

This is rather intriguing, because we have no idea how light works on the level of quanta, and even our knowledge about space is wanting, to say the least. There is practically not a single statement that we are a hundred percent confident about when it comes to space, let alone absolute and eternal truth. And yet we have built our entire measurement method on space. And let's not dismiss the other unsecure formula in the definition: time. Because traveling a given distance requires speed and space—and time plays a crucial role in these—we need it for the calculation. Yet our knowledge on time is first and foremost very inaccurate, and also extremely relative; and what is more, time exists as a notion only.

So one meter is calculated by self-justifying measuring units. These lack accurate, stable, and constant foundations. Maybe the platinum alloy was a lot safer standard than time, space, and light; at least we know much more about that.

I can't say I'm infinitely happy knowing that on our quest to infinity and coincidence we are armed with this battered and unreliable meter.

But let's return to the toothpick exercise. At the end of the process, when breaking the pieces in half was no longer possible, we were faced with a decision: We either keep looking for the origin of matter in the depths of infinity, or we stop and meet creation in the middle. If we take the first route, which basically claims that the reduction of matter goes on into infinity, then we agree with the definition of infinity as something without a beginning and an end. If we take the second route and say that the most elemental particle was created by a higher power, we either imply that infinity doesn't exist (because creation had a beginning, therefore a finite amount of time has passed since then), or else there's only one explanation: Infinity must come hand in hand with finiteness in order to actually mean something. The second case leads to believing that creation itself came from infinity, as it produces finiteness. This consequently makes *time* a component of *infinity*.

We can check this statement by asking if there's an instance when time is absent. We know one answer to this because we've already read about it in this book: As far as light is concerned, we can't talk of time. The notion of time is absent where the speed of light is involved. And where there's no time, there's no speed either. Still, I get to point B from point A, therefore it's possible to experience infinity. So, if the answer is light, then it means that infinity can't be defined when time is present; infinity doesn't contain time and speed, for instance, when it concerns space.

Not so long ago I said that time was a component of infinity. In light of the previous paragraph, I will modify it: One component of infinity is the complete *absence* of time.

What do we need in order to make infinity rational according to its own rules?

If you went with option B, creation as the answer to the toothpick-breaking exercise, then it only makes sense to suppose that infinity wouldn't exist without the illusion of time. Otherwise, finiteness wouldn't exist within infinity either. Finiteness functions within infinity: It can generate events delimited by time—we are not sure what time generates in infinity, but let's call them events. The variations of these events are further exponentiated, which is only possible if finiteness is contained in infinity—or else there wouldn't be enough room for every event generated by time. It's problematic, though, that time-infused finiteness is of no use when we want to understand timeless infinity—but we'll cover this issue in the chapter about dimensions.

The problem of the meter is pressing, but now we know a bit more. Meter or distance based upon speed loses its meaning in infinity it as infinity doesn't include time. We could still try using our platinum bar to measure infinity, but unfortunately, infinity contains all kinds of time, so all the stages of the platinum alloy's life would exist at the same time. It would at the same time mean the formation of its first atom, the standard bar that hasn't yet reached its full length, and the oxidation-stricken, decaying, useless object. All standards would be *the* standard.

So can we dismiss the idea of infinity? Or is there perhaps another way to understand it?

There's an ancient, more than two thousand years old, paradox. Let's take a look at its scientific solution, but brace yourself—it's not going to be any less shocking than the previous ones.

Let's imagine that we're walking on an asphalted road. Suddenly, there's a crack in front of us, spreading all the way to the sides. The crack is infinitely deep, so whatever we drop inside, it will never reach the ground. But it's only ten inches wide; we can step right over it and go on our way. This way we've encountered infinity, but didn't become part of it since we didn't fall in. When we're not part of infinity, it has no effect on us and we can observe it.

The next thought experiment gives some sort of solution to infinity. Suppose that we are going on a family vacation to the beach. The hotel we booked is three hundred miles away. That's a pretty long distance. It would take roughly eight to ten hours of driving altogether, and for safety's sake, we come up with a rule: to always drive only half the length of the remaining distance and then stop to rest. It's a simple enough rule that may get us to the beach safely, family and all.

We set off. Driving on the highway is a piece of cake, no delays, no traffic. The miles just fly by. Three hours pass and we've already made it through half of the journey. We may feel tired after driving 150 miles, so it's a good thing we came up with the rule to always stop halfway and rest for a while. We still have 150 miles to drive, but that's not so bad because we know that we only need to keep going until half of the rest of the way, and then we can have coffee and relax a bit. After this stop, we only have 75 miles to go to reach the sea. That doesn't seem much; it can be done without any more stops. But remember the rule: We must always stop halfway—and we keep our word.

The next stop comes after 37.5 miles, it takes less than thirty minutes. We rest, and sincerely feel like we've just stopped, so it would be better to continue the journey, but we made the rule, so after 18.75 miles, we stop again. We start to think: The next stop will be in 9.375 miles. Is it really such a good idea to keep following the rule and stop halfway? Well, we've made it this far, so we might as well finish what we've started.

There are a few more stops coming up at the following points:

- 4.6875 miles;
- 2.34375 miles;
- 1.171875 miles;
- 0.5859375 miles = 1031.25 yards;
- 515.625 yards;
- 257.8125 yards;
- 128.90625 yards;
- 64.453125 yards;
- 32.2265625 yards;
- 16.11328125 yards;
- 8.056640625 yards;
- 4.0283203125 yards;
- 2.01416015625 yards.

It is scary. We are now right in front of the hotel, two yards away from the entrance, and yet we can't get in, we're bound by the rule.

As we keep dividing the remaining distance in half, the thought occurs that we may never be able to enter. Because even though we're only inches away from the entrance, we still have to stop at the following points, according to the rule:

- 1.007080078125 yards = 36.2548828125 inches;
- 18.12744140625 inches;

- 9.063720703125 inches;
- 4.5318603515625 inches;
- 2.26593017578125 inches;
- 1.132965087890625 inches;
- 0.5664825439453125 inches.

This still feels like an eternity. Let's take the phone from our bag and calculate how many times we need to stop before we can finally enter the hotel. The result is embarrassing. We can *never* cross the threshold of the hotel. Because of complete divisibility, the distance is infinitely divisible. Zeno of Elea (488–430 BC), the Greek philosopher, encountered the very same problem a long time ago. It took us almost two thousand years to solve his paradox.

Let's unveil the solution!

On our way to the beach, our last stop was about half an inch away from the hotel door. Up to this point, we have stopped twenty-six times, and I'm convinced we are able to make it to the finish line following the rule. So let's go on at a deeper level of thought.

When we reach the 120th stop, the remaining distance is so short that it's actually the Planck length, which is 10^{-33} cm. Once we reach it we may throw classical physics out the window, time doesn't exist anymore. Planck length is the distance we make without speed, without even moving. We still change position. There is no transition between the two positions: One moment we're in one position, and in the next in another.

Physicist John Wheeler of Princeton University studied the phenomenon along the lines of Einstein's equations, which say that at the level of Planck length, space warps to such a degree that it's no longer flat but rather resembles a cylinder.

Of course, we don't experience this kind of movement or traveling in our everyday life. Hopefully, in the not too distant future, we'll be able to measure Planck length, but until then, we can prove it. Imagine an object that travels half an inch at a constant speed. While it's on its way, we take 10^{33} plus another few pictures of it in a row. The object would appear in 10^{33} pictures but would not be present in the few extra that were taken. It would be as if the object didn't even exist. We may see a blurry figure at most.

I have John Wheeler's theory at the back of my mind as I'm drinking my coffee, and I can't help but wonder how many jumps in space-time it takes until the cup actually reaches my mouth. If I dismiss his theory, I'm left with the question: How long is the shortest possible distance? What is the shortest distance? How does *moving* work?

In the related chapter, we defined time as a period we observe between two events. So whatever happens between two Planck length-long space-time jumps can be sensed as time passing. If my coffee cup is circa five inches away from my mouth, it takes at least 10^{34} separate movements for it to get there. Those 10^{34} events as a whole happened in zero time, yet there was a little bit of time passing between each particular movement. But why did it pass? How come we have to wait a little between taking two consecutive Planck lengths?

We've had a long journey, with an infinite amount of stops, and if John Wheeler's theory is right, infinity cannot be found because it's present in the curvature of space-time (it's the place where time doesn't exist). We have just used infinity to create finiteness.

2. To Quarks and Beyond

The chapters on light and on infinity provided a glimpse into the discovery of the elementary world, i.e., the further we go, the more phenomena we find that we do not understand and cannot explain. And when I say elementary, I mean the smallest elements we know now. Quarks and mesons are extremely small particles compared to even a single grain of sand—just as our planet is extremely small compared to the whole universe. It's like a single cell of a single person to the number of cells of the entire human population of the earth—that's how small the earth is against the backdrop of the observable universe. Looking at the numbers, that would be 1 cell compared to 50 trillion multiplied by 8 billion cells.

Inside an atom, there is a particle called meson. One of its elementary particles is the quark. Inside the universe, there is a planet called Earth. On Earth, there's a grain of sand. That single grain of sand contains atoms, as many of them as the number of seconds that have passed since the formation of our universe. And right there, in one of those atoms, we find the meson, with our tiny little quark inside.

Does it mean that everything is made up of atoms?

Physicists George Gamow, Edward Condon, and Ronald Wilfred Gurney studied the process of radioactive decay in the context of of the uncertainty principle that comes from quantum mechanics. They discovered that it is impossible to simultaneously define the exact position and direction of a given particle. This is the reason why a particle can go through a barrier as if through a tunnel. It's called quantum tunneling. There is only a slight possibility for it to occur, though. It is a process where we create conditions in which an electron can propagate through insulation material without making any lasting changes in the structure of the material itself.

Quantum tunneling is one of physics' crucial theories; its results are put to use even in certain electronic devices. In oversimplified terms, the phenomenon can be pictured like a ball that would never wear out plus a wall that would never crumble, and I can go kicking the ball at the wall, and there would be a time when the ball wouldn't bounce back. It would go through the wall in such a way that neither the ball nor the wall would show any signs of what just happened. Because of the mass of the ball and the number of its atoms, of course, I would have to kick it for quite some time before I could witness this.

It looks like almost any event that was ever predicted is possible on the level of quanta if we create just the right number of possibilities. Murphy's law states that whatever can go wrong will go wrong, and it applies to the quantum level as well. Whatever is possible, will eventually happen.

But what are quanta and quarks in the first place?

Physics lessons up to this day present particles as simple and point-like, even though we've long past that idea. String theory—and later on the membrane theory or M-theory—was built on the idea that the elementary component of an atom is an oscillating string, not a multitude of dot-like particles as was previously believed. Think of it as a kaleidoscope: the more you turn it around, the more magical shapes you can see. We only suppose that an electron is dot-like because we don't have an appropriate device to observe reality. But if we were able to see the electron, we would see a string inside. And it is this string's oscillation, frequency, and wavelength that determines which elementary particle we're going to see or measure. If we stretch out the string and strike it, it becomes a quark; and if we let it loose and strike it, it becomes a neutrino. What we see and measure is what we strike.

M-theory unifies multiple versions of string theory, practically, with the addition of membranes that can have 6, 7, 10, 11, or even 26 dimensions. String theory is like an orchestra: different instruments, each with its own frequency range, come together to create the whole masterpiece. You can play rock or lyrical, romantic music with the same instruments. An orchestra needs a composer, someone who puts together the melody and chooses the appropriate instruments with different frequency ranges, and someone who tunes them. When all of that is done, it's time for the conductor to step up and direct the orchestra in front of an eager audience. Well, string theory seems to hold the key to many unanswered questions—among them how there can be more than four dimensions—but we don't know who tunes and how; who is the composer and conductor of this great symphony.

Even a few hundred years ago, it was only natural to say that God was behind an event that was impossible to interpret. People went on to build entire theories on this foundation. However, in the last seventy years or so, it's discovered that many of those inexplicable phenomena had nothing to do with God's direct intervention. For example, the answer to the question of why it is dark at night even though the universe is infinite and there are an infinite number of stars, meaning there must be infinite light—as the Olbers' paradox says. Generations of astronomers and philosophers tackled the question and surrendered to the only possible solution: the paradox can only be solved by accepting divine intervention. It shouldn't surprise you if I said that things have stayed the same since then. We so often get to the root cause only to find that there's no rational explanation. One thing changed: we took the material apart into even smaller parts; we came up with new theories, each more complex than the one before. But we still don't know the basic entry points to the materials, into the theories. Even the most genius theories reach a level where we can ask, how did this come about; i.e., what was the cause of this particular effect? All those scientists who have reached the highest levels of understanding assume the presence of a higher power in most cases, one who devised and created everything. Some of them are brave enough to call that power God, such as Kepler, Galilei, Descartes, or Newton did. Others are less brave, and instead of calling it by its name, come up with another army of theories where the similar basic cause-and-effect question does not yet arise—but it will, whether five, a hundred, or even a thousand years later. Scientists are working hard to find the one unitary formula of creation theory that would explain everything from the beginning to where we are now, building itself on its own. As Einstein famously said to one of his students, "I want to know God's thoughts—the rest are details."

If I drop a glass, it will break into pieces when it reaches the ground. Is it possible for the glass to break into pieces before I drop it? So that it breaks first and then reaches the ground, the exact same glass? Let's look at the process from the perspective of logic, and see what decisions were made. The first decision was to let go of the glass; this caused the falling, the falling caused reaching the ground, and this resulted in glass pieces, as an effect of smashing. So we see that the first decision set the course for an entire chain of actions. That first decision is irrevocable: the glass falls, and this makes the effects calculable. In some cases, however, the order effect can overtake the cause. To be more precise, cause and effect can be in a relative environment, such where there's no time. To analyze this phenomenon, let's proceed with our double-slit experiment that we encountered in the chapter about light.

As the laser light was let through the two slits we made on the cardboard, the projection on the wall resulted in interferences. As we know, interference can only be caused by waves that meet each other. The experiment was repeated under laboratory conditions, letting one photon through the slits, and once it reached the target, the next particle was sent—the result was the same interference projection on the wall. The wave nature of light was caused by one particle. It contradicts the everyday experience; in our daily life a single particle cannot move like a wave—at least, it needs much more time. Various theories attempted to prove how a single particle is capable of interference, i.e., interfere with itself, but none of them can be tried and verified. All of these theories are built on the same premise of quantum mechanics regarding momentum and position: we cannot know bothvalues at the same time so that we can calculate any result based on them . If we know in which direction the particle moves, we can't know its position; and when we have its position, we cannot know where it goes, its momentum.

In order to reach a deeper understanding of the double-slit experiment, we'll take a short detour. You must have heard the adage that measuring affects the result. Let's look at this first. Imagine a cold winter night when it's 14°F outside. Inside it's nicely warm, 79°F. I put a test tube on the table, which is a few yards away from the fireplace. I fill the tube with 104°F hot water, and I put in a thermometer cooled to 14°F. After a couple of minutes, I go check on it. I find that instead of 104°F, it shows only 72°F. The measuring device affected the subject of measurement, i.e., the thermometer cooled down the hot water in the test tube.

It's a perfectly common phenomenon in our daily lives. But if I want to present the conditions of quantum mechanics in the same experiment, it looks different. I insert the 104°F hot thermometer in the test tube with the 104°F hot water. They have the same temperature. Once I drop the thermometer in the water, I immediately leave the room and go outside. I can at once feel the wintry coldness of the 14°F out there; it's freezing. I take a walk on the snowed pavement, right to the window, from where I can see the table beside the fireplace. If I look close enough, I can discern what the thermometer shows, which is 72°F. Because I can't believe my eyes, I go back inside. The 104°F hot thermometer put into 104°F hot water should show a 104°F temperature. Once I'm in the room, I check the mercury line of the thermometer closely—and now it shows 104°F. Come to think of it, this is how things are in the fourth dimension in the realm of quanta, and this is how observation changed the result of measuring.

If we look at the results of the two thought experiments, we find only one difference between them. In the first one, we know what happened, how and why, while in the second we don't. We can't even declare that the phenomenon exists or that it doesn't. We don't know what it is about, let alone how it works.

We're getting deeper and deeper in understanding. It's time to add the final touches to the double-slit experiment.

John Wheeler came up with an idea that he called "delayed-choice experiment." It is similar to the double-slit experiment. There is one particle that starts moving toward the slits, and when it's gone through the cardboard but before it reaches the screen, right at the last second, we make a choice. Do we want to observe the trajectory of the particle or not? That is, do we want to observe whether it has crossed both slits at the same time (in which case there will be interference) or whether it hasn't (in which case there will be no interference)? John Wheeler and many others after him found that should we want to observe the trajectory of the particle (i.e., did it cross both slits), it will cross one slit as a single particle and we won't see any interference. But if we don't want to know whether it crossed both slits at the same time, the particle retroactively turns into a wave and we see interference.

Cause and effect switched places. First there was the effect, and then the cause.

The trajectory of the particle, its own *past*, was defined *later*, according to our dimension. We decided on measuring only after the photon crossed the slit(s), when we thought it made the choice to go through only one of them or both. It's as if our decision whether or not to measure made the photon go back in time and choose its trajectory accordingly.

So the answer to the question of the broken glass is here. On the level of quanta, it is possible for the glass to break before it reaches the ground. Or it can break before I decide whether I'll reach after it or not.

There are a number of theories to verify the double-slit experiment, but none of them can be proved. We must admit that we don't know how it works; still, the phenomenon exists all the same.

There's one thing we can safely declare. The world doesn't work the way we experience it day by day. We've seen that time doesn't exist for real, even though we need it for speed and motion. Let's try and get used to the idea that the sequence of things—the order in events, one based on another—only exists in our minds. And if we seem to encounter something inexplicable that looks like a miracle, we have the choice to be humble instead of hostile. We can stop doubting. Now we know that when we don't understand something, it can still be a thing that exists and is worth exploring. It may not be the right way to stay comfortable and choose not to know what's behind it. The way to development is tostay curious and humble, and accept our mistakes whenever we make them.

3. The 4+1 Dimensions and Beyond

Dimension can be referred to as size or spread. We use it as an index, as a means that helps us find or realize things. For example, take four steps to the right, and then six forward. We need as many dimensions as there are independent values when we define a point or an event. This sounds easy but already predicts that we will run into difficulties, because we'll have to find independent data types. Therefore, there can be cases when we might not even know what exactly we're looking for. One such case is when time is absent; we will then need something to substitute time. Or maybe we don't?

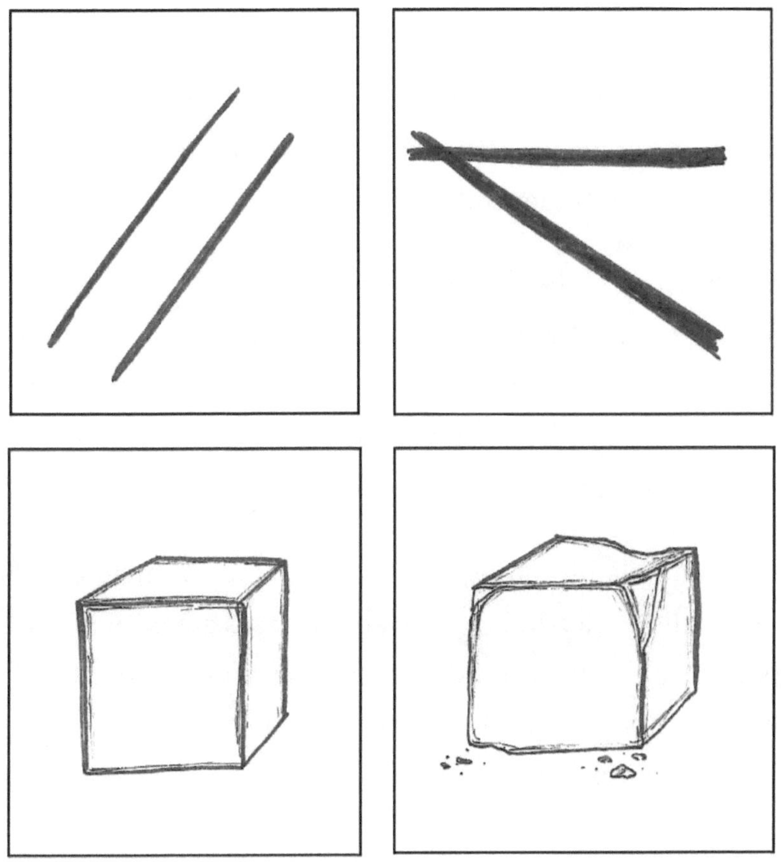

3.1 Understanding Dimensions

It's easy to imagine and understand four dimensions. The first dimension (1D) is simply a line or a number of parallel lines on a plane that never meet. When lines do intersect, there is width and length, this is the second dimension (2D). In the third dimension, there's width, length, and also depth. The objects forming our environment are three-dimensional (3D). The fourth dimension (4D) is time. According to our definition of a dimension, we can state that we can describe our existence and the environment which surrounds us with four independent variables.

In each dimension, the one dimension lower is always affected by the one higher. And not only affected, but it's crucial in order to understand and create the dimension one higher. Two or more 1D lines make up a 2D shape, and several of these are used to form a 3D object. And with the addition of time to 3D, we can create 4D.

Take the example of putting a book on a table. A single page of a book is two-dimensional, but once all the pages are printed and put together, we have three dimensions and our book is created. The fourth dimension is time. Let's grab the book and place it on the table. The book moved in the three-dimensional space, using the fourth dimension.

The example above is not absolutely true. In reality, dimensions are not exactly created like that. For instance, since there is no depth in 2D; we can put as many 2D shapes on top of each other as we want, the best we can hope for is a seemingly random two-dimensional mess. In our example, we had to print out the 2D pages and so *make* them three-dimensional *ab ovo*, in order to create a three-dimensional book. In reality, putting multiple 2D shapes together is more like putting several transparent digital photos or pictures on each other. See the example below, which is made of thirteen different 2D pictures:

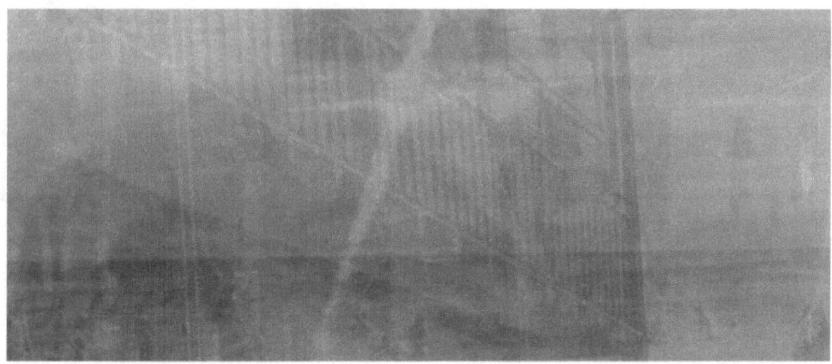

It is also impossible to create an exclusively three-dimensional object because any 3D object is created in space and becomes part of space—and space always includes time, so we just can't get around the fourth dimension. The moment something gets into space, 4D will immediately be a part of it. Imagine a steel cube in the making. It gets connected to 4D as steel is being formed and as soon as it's finished, millions of atoms are already breaking off from the cube. Even while you were reading the sentence above, the size of the steel cube already changed. It changed, albeit minimally, because it encountered time. 3D objects only exist in our minds.

3D as we think we know it works in theory and on paper; we can make calculations with it. It doesn't contradict the times tables, as Einstein once remarked. Still, we can't perceive it. What we perceive is merely an illusion. One and two dimensions are also rather theoretical notions; we can use them for calculations but can't put them into practice.

The zero dimension contains the geometric point that we have mentioned in the chapter on infinity. There we said, it has "no length, area, volume, or any other dimensional attribute. It's basically a zero-dimensional location." We have agreed that we can't possibly define a straight-line segment using the point as zero dimension. If we applied the definition literally, it would mean that the line segment starts and ends at zero, i.e., at nothing. But there's no logic in that. Even when it comes to the formation of the universe, we suppose that it started from *something*—energy or mass perhaps—but not *nothing*.

Most of us think that because we can use zero, it's fine to use one, two, and three dimensions the way we do, because the calculations with them are correct. It doesn't seem to be a problem when we depict two dimensions as the page of a book, although we know that a sheet of paper has thickness, and therefore it's not even three-dimensional but, by rule, four-dimensional. We're resourceful; we can start with incorrect facts or data in a way that leads us to a correct answer. We make do with what we have. We accept the mutual agreements on one, two, three, and four dimensions and continue to work with these, stopping only when we run into significant contradictions.

3.2 The Future Is Yours and the Past Is Mine
—Or the Past Is Yours and the Future Is Mine

It's now 2020, and all the talk is about COVID-19. We're depressed and frustrated. We have no idea what's in store for us. Will it hit my vicinity? How will it affect my life? Will I get sick? If only we could see the future!

Is there a way to look into our future—or even our past, for that matter? Or is it all just a never-ending *now*?

What we experience at the level of existence is the *now*. We do believe that the past used to be *now* once too, and we can only hope that the future will turn into *now* as well. Does that mean that dimensions may contain relativity? Maybe so, and maybe relativity is one of the extra dimensions.

Let's place ourselves in our own dimension.

Right now while our house is being refurbished, I'm writing this book in the living room, which used to be the TV room. When I look up, I see the oven in front of me and the dining table next to it. The fridge is behind me. They are all supposed to be in the kitchen. Everything is covered in dust, including me. My laptop would need cleaning every hour if I wanted to keep up with it. But why would I bother? This is what renovation means and the builders are everywhere. I experience the present.

As you're reading the book, the builders have long finished their job on my house, and I can enjoy the comfort of our living room once again. Still, as you read these lines in my future, you caught a glimpse of my *now* — of that very state and moment in which I exist at present. It seems I'm sending information from the present to the future. As you read these lines, from your perspective they will contain past information. There don't seem to be any complications with this: from my viewpoint, interfering with future works. I create a book that you read in my future.

Well, another day has passed since writing the last two paragraphs; I wasn't too productive last evening. I have just re-read them now so that I could go on, and I've stumbled upon something amazing. Those two paragraphs are about my yesterday's present, and that means it's already in the past; but with you — the reader — in mind, it's still information directed to the future.

If you skipped over the last few paragraphs about my renovation, I wonder what the information I gave there would mean to you to-day: would you consider them future or past?

That's a bit difficult. Let's try a different approach.

I would like to ask you to go through the next five pages slowly, considering them carefully.

1

2

3

4

5

Now let's see what happened during these few seconds in my dimension and in yours.

Over here, in my dimension, I'm still sitting at my laptop covered with dust, writing this book. Nothing significant has changed in the living room, although the builders have made some progress even in these five seconds.

But I did something remarkable. I seriously affected the lives of millions of people, and I did it from your perspective, I changed the future through the past. In my own present, I asked you to stop—from my perspective in the future—all ongoing processes for five seconds.

For five seconds, while you leafed through the last five pages, the future was exactly as I, while in the past, wanted it to be. And from now, the events in your life will happen with a five-second shift, i.e., five seconds later than they would have happened if I hadn't asked you to study those five pages. This five-second delay has been caused by the combination of my present—my *now*—and your past, defining some parts of your future and also some of mine.

In your future, these five seconds may prove enough for you to avoid an accident, although you may just miss the bus or train by five seconds, too. This five-second delay will affect everything you do and will pass on to everyone you meet for the rest of your life, and they will transfer it to further people. And so, it will spread and multiply.

Wondering why? Well, as you were reading the book, your daily life was going on. You would have done something else but you didn't. Still, you must do it some other time, causing the delay of that thing that you would have done in *those* five seconds, and it goes on and on. You may say that you weren't supposed to do anything in those five seconds anyways. Then I'm afraid you don't see the parallel dimensions yet. Why don't you experience this event delay? Because you don't have a point of reference, you don't know what would have happened without the five-second delay. And as the minutes and hours go by, those five seconds will seem less and less significant. It will become evident that they are a part of your life from now on. In two days, you won't even remember the experiment, and when something unexpected occurs you may simply call it coincidence.

From now on, be on the lookout for those five seconds!

In my *now*, I can't do anything against the five-second event delay, because whatever I'd suggest would come from me and my present, affecting your present and future even more. It's up to you to stop the event delay, although the book is here to guide you.

Wait—in fact, there is one way I could make it all go away: I would only have to delete the exercise.

Would you like me to do that?

Unfortunately, I can't hear you here in my present. That's again an interesting phenomenon, but let's just conclude that in our four-dimensional world, we can initiate certain events and chains of events in space-time. From my present state, I can seriously affect the future.

From my perspective, I changed your future from my present.

From your perspective, I changed your future through the past.

Here, in my *now*, I cannot know what effects that five-second event delay will bring to your life. If I knew *now* that those five seconds would save people from a mass accident, I'd most certainly delete the entire chapter. But all I know is that I've started a chain of events to unfold in the future, and I have no idea of the conse-quences. If you ever, in a faraway place, run into a friend whom you haven't seen in a long-long time, remember: it's not a coinci-dence. It's because of those five seconds. That's all it took for you not to meet, but you did.

It all makes sense; still, it doesn't seem to resonate with us. Now, why?

Because it lacks cause-effect that we're used to. We live in a world with fast and strong effects. If a ball hits the ground, I expect it to bounce back immediately. If I exceed the speed limit and run into the patrol, I know that I will have to pay up in a week or so. It won't surprise me; I am expecting it. Whenever we pollute the environment, we are aware that there will be a time when the realization hits us hard, but at such a faraway point in space-time that we can hardly imagine. We believe that the particular effect will not affect us so we might as well forget about it. And yet, at one point in the future (which will be someone's present), environmental pollution will get unleashed upon us, producing more and more peculiar bacteria and compounds that we will believe to be a coincidental consequence.

Every cause and effect concerns time. But if this effect takes place in a space-time that is outside our perception, i.e., we don't experience it within a reasonable time period, for some reason, we tend to reject the idea that it's the effect of a cause. We usually put it down to pure chance, without any reason.

By this five-second event delay, I initiated unforeseeable events in the future on purpose, but those who will ultimately experience its consequences will consider them coincidence—because they don't know what caused it.

This five-second event delay makes it clear that consciousness affects events—let's call it an event index. The experiment adds a lot of value, pointing out that the more accurately I can initiate future events in my *now*, the more conscious I become about the possibility of doing so. Of course, I can't know for sure how the experiment would play out, but I am conscious of the fact that I've caused an event delay on purpose.

To summarize our findings on forming the future, we can state that it is possible to manipulate future events from the present. And whether the manipulated person will notice the consequences or only the manipulator, depends on the manipulator's level of awareness. On the manipulated person's side, it may seem like a coincidence. If the manipulator and the manipulated person is one and the same, his perception level will depend on the weight of his awareness, i.e., to what extent the manipulation was directed.

Between past and future—or present and past, looking at it from another perspective—we have found the necessary index, or rather, dimension. It is information.

What must happen and when? If I know *now* that something will have an effect in the future, and I'm able to carry this information through time until the target date, then this particular event happening at the given time in the targeted future will not take me by surprise. But if I didn't carry the information that connects past and present, I will perceive the event as pure coincidence when it happens in the targeted future, although it was initiated in the past.

Maybe this realization can help us solve the paradox I mentioned in the chapter on time: *if you do something today, how will you prove tomorrow that you have caused it the day before?*

For the time being, there is only one way to prove that the work I accomplish today affects tomorrow: the work itself is the information that is carried on. If the job is done, i.e., the information is destroyed before it reaches the next day, then there's no way to prove that I've accomplished it the day before.

So, information is one of the possible links between past and future. And if information can be sent from the past into the future, then the future is predictable; there can't be any "coincidences" that can't be explained.

Just think about it. We manipulate the future on a daily basis—although not on purpose. I say that I'll meet a friend the next day at 11 am. I promise my wife to call her in the afternoon. I automatically create a piece of possible future from the present. All eight billion of us are doing it, thousands of times a day, and the consequences of these events are thousand times thousands. And the more conscious we are about creating the possible future, the better we understand the present of that consciously created future. Literally speaking, we create the future. And it starts in our mind—that's the creative power of the mind.

Now, let's see why we need the fourth dimension. What is its purpose? How would the world be without it? Without time, there would be immediate access to all information from any one place. We would be able to see the outcome of each possible event that we have caused, so we could easily pick the one that we liked best. And because it wouldn't be enough to know only the event possibilities that concern us—because they may be connected to other people's event possibilities—we would need all information of all possibilities of the whole mankind at once, at the same time.

Let's summarize how the timeless dimension would be for us, humans. Everyone would be aware of the future consequences of his actions and thoughts. And everyone would know about everyone else's actions and thoughts and their possible consequences. Since nobody would want to live a useless life, everybody would pick options with consequences beneficial to themselves. We'd have an infinite number of possible events; uncountably infinite, because each possibility springs from another. Mankind could choose an infinite number of events with beneficial outcomes for each single person—therefore, it would be useful for everyone else as a whole. This is the dimension without time made understandable.

Now let's fill time back into this timeless dimension, to understand the illusion of time. In a dimension without time, there's no possibility to experience, to endure. There's only omniscient information, a state of everything happening simultaneously. Only a dimension including time can offer the chance to experience and live certain events, the possibility to know or not know certain things.

We shouldn't believe that the timeless dimension is impossible. Impossible only proves that we have so little knowledge. We're in the age of quantum computing, looking for uncountably infinite sets in countably infinite sets. We must analyze the uncountably infinite set of possible events initiated by a single person and add it to the countably infinite set, humanity itself. And since time is absent at the level of quanta, infinity can be understood and handled. We have to wait only a little longer and understand some more until the first, real quantum computers can be built. Once we have them, each and every possible event can be accessed, and all we have to do is trace them. We'll get a map to the future.

We've seen the link between past and future that starts from the past and affects the future. But do past and future exist? In this chapter, we've caught a glance of the past—or to be precise, you have seen my past, while reading this book. That still is the present for me, whereas in my present you belong to the future. You may say that it's just a twist from the writer, and it only aims to muddle definitions, thus keeping back information and creating a paradox. Very well, let's find further proof, approaching the subject from the other side. Let's see where we go. Can we see the future?

That's not a good question. Whose future is it? But let's leave it for now and start with the statement that future cannot exist without past. Past can very well exist on its own as long as no information is carried into the future. That's a significant contradiction, and it also contradicts the direction of time, i.e., time can only go forward. Exploring the future, which is what we are about to do, needs even more dedication, attention, and understanding. We'll have to be stricter in analyzing contradictions and only accept a theory if we don't find a logical way to prove it wrong.

It's early in the morning. I'm in my car in the city, on my way to work, stuck in traffic. It's these annoying traffic lights, they cause the endless queues of cars. But come to think of it, without them it may be even worse. I'm waiting at the red light, counting down the seconds, because I know when it will turn green. Does that mean that I can see the future? No. I have no idea what the future exactly holds. The traffic light might break down, or someone might crash into it while I'm patiently waiting in my car. All I can do is assume events of a tiny segment of the possible future; I consider the most frequently experienced events to be certain to happen in the future. And this means that since it always took fifty seconds for this particular traffic light to turn green, it will do so in this instance as well.

Still, what if someone does take the wrong turn and crashes into the traffic light? Considering the traffic light, it may look like sheer coincidence. But from my perspective, this already played out as a possible future event, and now the slight possibility became 100%; it did happen. But let's just go back to the traffic light for a moment. Was it really a coincidence that the car crashed into it? No, it wasn't. The chance of a car crashing into the traffic light was in its possible future all along. The traffic light stands at a road with cars rushing by at 30-50 miles per hour. If the traffic engineer had made a decision in the past to install an overhead traffic light instead, then he would've changed all the possible and probable outcomes. The traffic light would be impossible to be crashed into. That one decision from the past would have an effect on the past, present, and on the future as well.

The cause-effect relationship between past and future is somewhat muddled in this example. Nonetheless, the cause is still in the past. If at one point in the future a decision is made to remove the traffic light at the side of the road, then further accidents can be avoided, i.e., no cars can crash into it. But if the cause is in the future, it can't affect the past. So even if a car crashes into the traffic light in the future, it cannot make the engineer of the past design the lights to hang overhead.

The example may seem forced, and it's because of time and relativity. But analyzing what we've seen so far, we can state the following. If I create information in the present, then as soon it is used—in the future—it becomes past. It's as if time turned around, if we look at the perspective of information itself. Past is no longer the starting point. Based on this logic, an independent observer would describe the process of time like this: time flows from present toward future and from future toward past.

3.3 Stars and Time

The next exercise will bring us closer to solving the mystery of future and past. But before we start it, let's take a look at the definition of illusion. Wikipedia offers the following text:

"An illusion is a distortion of the senses, which can reveal how the human brain normally organizes and interprets sensory stimulation. Although illusions distort our perception of reality, they are generally shared by most people. Illusions may occur with any of the human senses, but visual illusions (optical illusions) are the best-known and understood. The emphasis on visual illusions occurs because vision often dominates the other senses. For example, individuals watching a ventriloquist will perceive the voice is coming from the dummy since they are able to see the dummy mouth the words."

We can distinguish between different types of illusions, such as visual, auditory, tactile, and temporal illusions. Experimental psychology often explores the unique ways in which we receive these illusions, to get to know the way of human perception. Oftentimes, even knowing the inner workings of an intended illusion can't convince us of the truth behind it—that it's an illusion.

Imagine it's a warm and quiet summer evening. Taking a walk along the seashore, lying down in the sand, and gazing at the stars is pure bliss. It may be the most relaxing way to spend time. And the longer we stare at the night sky, the more stars we'll notice.

There's the Andromeda Galaxy, for example. It is about 2.5 million light-years away, meaning that a ray of light sent from the earth toward the Andromeda Galaxy would reach it in 2.5 million years. But how long is that period? We can imagine it better if we think about Europe and that 2.5 million years ago the continent was covered by ice 1.5–2 miles deep, devoid of human life, with mammoth herds roaming the surface—according to Darwin.

And now think about those light particles that started at that very moment from Earth and how they reach the Andromeda Galaxy right now—on February 2nd, 2020, at 4 pm—hitting their telescope. The inhabitants of the Andromeda Galaxy analyze what they see: strange four-legged creatures are killing each other; others are digging into what appears to be a huge, white cover and pull out green things from underneath. But the inhabitants of the Andromeda Galaxy are completely mistaken. On February 2nd, 2020, at 4 pm, about eight billion people live on Earth, and those strange four-legged creatures who used to go about killing each other are extinct. (Nowadays it's rather the two-legged creatures who go about killing each other.) So, the Andromedans couldn't be more mistaken, their basic knowledge is simply wrong.

Can we agree on this?

It's not an easy decision. What happened is that the information about our past has just reached their present. Their truth is completely different to ours—yet both of us are right. What do we do—what do we spend billions of money on—when we explore space to look for habitable planets? We're combing through the past. We have all heard about newly discovered habitable planets—only millions or billions of light-years away. But we only know that those particular planets used to be habitable that many years ago. The planet that looks uninhabitable today, might be inhabitable in their *now*—only we didn't receive *that* information yet.

There's an interesting fact regarding these faraway planets. Our *now* is in their future. And their *now* is also the future for us. Suppose that we received evidence of an intelligent life form from the Andromeda Galaxy. We can be sure that those intelligent creatures have long destroyed their civilization by now—if they are a bit like mankind. It would take us 2.5 million years to witness that reality that we believe to be true on Andromeda. Whatever we know about the Andromeda Galaxy is 2.5-million-year-old information—*it's been obsolete for the last 2.5 million years*. There's about fifty percent chance that the Galaxy doesn't even exist anymore because of some unknown effect which occurs once in every five million years. And there's just as much chance that when we wake up to-morrow, we encounter a 2.5-million-year-old super-intelligent ex-traterrestrial species flying our skies.

So we can state that when we observe a distant planet, we perceive its past. And when someone observes us from a faraway planet, they also see our past. Both of us are thinking that we've perceived reali-ty—but both of us are mistaken. What we think to be true is false, simply an illusion. If we know that the information exchanged be-tween these two planets is based on illusion and we see the past, then we can state that these smart creatures of these two planets are aware that should they want to perceive the actual state of the other orb and they have the necessary device, then what they'll see is their future. From their perspective, they are analyzing the future, and regarding their own relativity, they can see into the future.

My plan was to go on and discuss the expansion of the universe, but now I see that the expressions *perception* and *observation* keep turning up. We should clarify their meaning before we move on.

We touched upon the notion of perception in the chapter on light. We know that we can perceive something if we can counteract the result of an interaction, i.e., when we see, hear etc. as something happens to a thing or a creature. But in several cases, we don't even know what we should perceive in the first place. Take, for instance, baby rattles. If we give one to a small baby, the baby will wave it around, listening to where the rattling sound is coming from. If we give it to a toddler, as he waves the toy, he'll notice that there's something inside the rattle and that is what makes the sound, even though he's not strong enough to break it. And if you give the same toy to a bigger child, after a shake, he'll take it apart at once because he knows that something inside is making the rattling sound, and he wants to know how it works. As a result, he'll discover that there were small balls inside and the sound was coming from them clinking against the wall of the rattle.

There are still some tribes all over the world who don't wish to be contacted by civilization. Because we haven't been able to establish contact with them, we have very little knowledge of how the particular tribe perceived our attempts to do so. If one of those people were willing to experience our world, we would get interesting feedback. If I gave him a small AM/FM radio, asking him to tell me as many things about it as he can, where would he even begin? At first, he would probably observe the exterior of the object and state its material is unknown to him, as he's never seen plastic before. He might as well smell it, touch it, taste it. Maybe he would try how it performs when using it to crack open a nut. At one point, he would inevitably push the "on" button, maybe by accident, and a voice would be coming from the small device. He'd sure be confused and scared! He wouldn't understand how this happened. How can this box talk? Surely there's no way a person could fit inside!

I'd assure him that it's not a trick of the gods, nor it is the voice of the ghosts of long-gone chiefs. The box would still be talking, and our volunteer would still be eager to explore. He'd take a big rock, and crack the device open. He'd find the strangest of bits and pieces, wires, and whatnot inside. And it would keep talking. Bewildered and annoyed, our man would tear the net-like mass of wires and the gadget would go quiet. If I asked what he thought about the radio, he would most likely answer that he couldn't find out how it worked, because right before he was finding the source of the sound, it stopped talking.

Our isolated volunteer has lived in another dimension. From this dimension in which he lives, it's not possible to discover the dimension where the radio talked. There is no information that links the two dimensions together, that makes it possible to comprehend the instances of the higher dimension. The lower dimension lacks the information that makes it possible to understand the higher one. There was no way our volunteer could've observed that the radio simply received invisible rays—electrons interact with these waves, then after a complex vibration process, the membranes of the loudspeaker start vibrating and we perceive them as sound.

And now let's take a real-life, present-day example, and try to decide for ourselves what the reality of past, present, and future is. The other day the following news was on TV: "In approximately sixty-three years, a spectacular astronomical phenomenon can be viewed for several nights, astronomer Laszlo Kiss said on Saturday on National Television. He expects the V Sagittae binary star will collide and shine bright, becoming a nova. It will become the brightest star for a couple of months in 2083, shining even brighter than Sirius."[19]

Laszlo Kiss was right and wrong at the same time. You might already suspect why I think so. This piece of news uses creative journalism to bring the future to us. Indeed, we are likely to witness something spectacular in 2083, but the actual text contains errors. For one, it talks about the event in future tense, although the V Sagittae binary star system is 7,800 light-years away, so saying it *will* collide is wrong. The collision, if it ever happened, has already happened. It went nova 7,800 minus 63, i.e., 7,737 years ago. So 7,863 years had to pass for the possibility of this phenomenon to become visible in our dimension. In 2083 we will witness how the V Sagittae went nova 7,863 years ago. Mankind perceives this event as the future, although it's the past. And when we get to the year 2083, which we then consider our present, and some time passes, we'll see it as past, too. The only difference being that by then the

---

[19] M1 News on February 1st, 2020, 15:36 p.m.

event can be considered past from both viewpoints, ours and that of V Sagittae, even though with a time lap of 7,800 years.

Between today and the year 2083, past and future coexist right until the collision happens. Does that mean that the adventure regarding V Sagittae will be nothing but an illusion in 2083? Well, it does and it doesn't. There is no way for us to witness the reality, the present of V Sagittae. We will see its past in our own present. And if by any chance V Sagittae going nova has some unforeseen radiation effects, we may experience the past of the star very intensively in the *now* of 2083. Of course, if we had some sort of sensor that would observe the star with twice the speed of light, then by 2050–2051 we would be able to receive the information of the event. This would leave us with at least thirty years to prepare for the event. Talk about time travel, right?

Do remember this example whenever you hear that there's no intelligent life in the universe beyond the earth since we haven't found any signs of it yet. Have your answer ready: true, such species might not have existed one or two billion years ago, but since then, one or two billion years have passed that we have absolutely no information about.

Do you think it wise and useful to plan to harvest crops tomorrow, based solely on the fact that it didn't rain yesterday? Useful it could be; we'd sure learn that we shouldn't plan like this. We can't build future plans on the assumption that the past keeps repeating itself. Maybe I shouldn't make any plans to harvest at all, until I find a way to predict with certainty that the weather will be favorable for this activity the following day.

3.4 Dimension Crossing

We have thus far perceived 4D. Let's half it and experience two-dimensional perception. Let's see how limited our observational possibilities are.

Imagine that we're out in a green field, a purely two-dimensional one. It has length and width, but no height or depth. It's completely flat. Even the grass grows in a way that it doesn't reach up toward the sky; it lies on the plane, horizontally. We can't see the source of light, because we can't see in space, it is invisible to us. Yet we perceive it, we experience light and darkness. Let's place ourselves in this two-dimensional world. It might seem difficult, but don't give up!

As soon as we're part of the plane, we only perceive the width of objects. A ball we kick looks more like a hockey puck, it's completely flat, it has no height. If someone approaches us, we can avoid the meeting just as we do in 4D. In this 2D world, we soon get used to recognizing objects. Our minds correct the sight, just as they do when we put on refraction-turning glasses that flip everything upside down. It soon turns the images of the world right back. We quickly get to the point when we easily recognize everything on our plane.

But what happens when the 4D world suddenly enters this wonderful 2D field? For example, a helicopter lands with a team of explorers.[20] We would be walking the grass field in our two dimensions, and notice two parallel, ski-like lines. Those would be the landing skids of the helicopter. From our perspective, they'd appear out of nowhere. Then we'd see some ovally deformed rectangles: those would be the explorers' shoes, as the explorers would get out of the aircraft, again appearing out of thin air. We wouldn't understand the miracle. How could this shape keep appearing and reappearing? The explorers are walking around. But we, placed in 2D, wouldn't have complete insight into 3D and 4D. All events taking place there would leave signs in our 2D world, but we wouldn't be able to decide where these come from. And because we wouldn't have the source information, we wouldn't know that the source should be found. We'd have absolutely no idea where to look for it.

We're having an unusual experience. We're becoming part of a higher dimension, but we don't know about this. It's affecting us but we're looking for its cause in our own dimension. Looking is

---

[20] The episode *Second Dimension* (*Második dimenzió*) from the Hungarian animation series called *The Mézga Family* (*Mézga család*) depicts 2D merging with 4D brilliantly, way ahead of its time.

difficult when we have no idea what it is that we don't know. We're at a tough spot right now because we're asking whether it's possible to see from one dimension into a different one and retrieve source information from there.

If we believe in this possibility, then we need to presume that systems with higher dimensions always encompass the lower ones. This would indeed enable us to look into lower dimensions and gather the necessary information. It would be true that it would be impossible to inspect the higher dimensional structure from a lower one. Information on it would remain unattainable. This logical conclusion is disappointing. We've seen that when we put multiple 2D objects on top of each other, no matter how many times, they never add up to 3D, therefore 2D cannot contain 3D. I'm not going further into this logic, as it wouldn't make the picture clearer—because of the complexity of 4D and our ability to understand such theories. The point is, crossing dimensions is out of the question if we presume that a low-dimensional structure is part of a high-dimensional one.

But if I don't presume that a low-dimensional structure is an obligatory part of a high-dimensional one, then dimensions can be crossed. In such a case, however, our basic theory on 4D is wrong, namely, that the lower dimension is a building block of the higher one. A third option would come in handy, as both conclusions are severe. Maybe we can settle on closure—albeit reluctantly—with the following statement: it is possible to see and understand the lower dimension from a higher dimension.

3.5 Parapsychology in a Nutshell

Parapsychology is an actual field of science, regarding observation. Its studies present irrefutable data. It's been said many times that what we don't perceive doesn't exist. But whenever we considered something impossible, it was always proven that we didn't know enough about it. So let's not make the same mistake and misjudge parapsychology!

Even though we can't consider ourselves experts in quantum mechanics, by now we have some insight into some of its ingenious effects. Even the human body is made of matter, and our elementary particles function according to the rules of quanta. What's more, as we move about, one Planck length at a time, billions of space-time jumps happen to our bodies.

We have very little knowledge of the human brain, and we can make this general statement about the process of how thoughts are formed. What we seem to know is the mass of conclusions drawn based on the results of interactions.

Quantum mechanics can be found in nature in many instances. It's there in photosynthesis or in the electromagnetic waves made by the brain. We encounter the absence of time in quantum theories as well as in our dreams. We can multitask: we walk, think, talk, breathe, maintain blood pressure, regulate temperature—all at the same time. We are able to switch the order of cause and effect: first, we think about the result of an action, and only then do we realize it. Does this ring a bell? Neuroscience makes use of quantum mechanical theories, as it's running out of rational ideas.

Even the most advanced supercomputer is incapable of true parallelization. Take the following task: there is a forest of unknown shape and size, full of trees. We need to find the shortest possible path from the edge to the pine tree right in the middle of the forest. An ordinary computer would start from the edge of the forest, and take each possible route one after the other, taking notice of the paths tried, as well as their length. It wouldn't explore the same route twice. Before it started each path, it would check whether that path had already been tried.

As opposed to this, a supercomputer wouldn't take one route at a time. It would simultaneously explore ten, one hundred, one billion paths. That's splendid—but what if the forest is enormously huge and there are millions more possible paths leading to the pine tree than the most modern supercomputer can trace? It would not be able to realize all tasks parallelly, and waiting for CPU time would start.

As we dig a little deeper into the soul of supercomputers, we find that parallel functioning is impossible *ab ovo*. Information is stored in the memory; it needs accurate labeling. Finding the right label, storing the data, and eventually retrieving it is time-consuming. Besides, the different steps of a given process cannot be interchanged, so waiting for one thread means waiting for the one that follows it, too.

A true quantum computer would solve the path-finding problem very simply. It would track all the possible paths that lead to the middle of the forest, and bearing in mind that time is absent at the level of quanta, the computer would be able to do this in only one way: at the same time.

I find too many similarities between the human brain, thought, or soul if you will, and quantum mechanics. Scientists are at a loss when it comes to premonitions, telepathy, or information gained through meditation. Yet extra-sensory perception (ESP) is real, you may have experienced it personally, too. It has a large history and quite the bibliography. ESP is not some kind of superpower of some higher elite race, neither is it magic. It's an ability we can all acquire with a bit or a lot of practice. Many of us practice it without even realizing. The spectrum of our everyday ESP experiments is quite broad: starting from throwing the dice and expecting some kind of result, to thinking of someone faraway and projecting a thought toward them. During the Cold War, entire research projects were dedicated to collecting information on the other party (Soviets/Americans), using remote sensing. And while absolute proof of results is scarce, we are aware that neither of these two nations is in the habit of throwing millions of dollars or rubles away on an experiment that is doomed from the start. Besides, there wouldn't be dozens of books and documentaries on this topic — check Google for Cold War ESP.

It often happens that when we call a friend whom we haven't seen in a long time, he picks up the phone saying, "I've been just thinking about you" or "I was thinking about calling you just the other day." It happened to me multiple times that I hesitated to do something for no other reason than that hunch. And ninety percent of the time it turned out to be trustworthy. Some may call it a self-fulfilling prophecy, but as far as the result is concerned, the label doesn't really matter. The takeaway is clear: when we have these premonitions, we don't know exactly what that future negative event would be, but when it happens, we can immediately tie it to that sensation. That's when we say things such as, "I should've just listened to my gut feeling."

And the list could go on. I'm sure you also experience similarly strange, irrational events that are based on premonitions. This proves that ESP exists. And it can be improved. I think we can differentiate between two types of people: those who haven't started using their ESP abilities consciously, and those who have stopped using them because they don't need them anymore.

What we can state as a universal truth may be: the phenomenon exists. Denying won't make it nonexistent. It's better to accept that we don't know something than to prove how little we know.

We can't seem to find any rational explanation to ESP within the limits of our four dimensions, so it only makes sense not to press it. Instead, we find a strange resemblance between ESP and the world of quanta that can help us understand ESP. One of them is cause and effect switching places or having certain sensations about the future. Analyzing parapsychology on a quantum level, we find that these events are not only likely to happen, but they must happen exactly as our gut feeling has shown it.

We began by looking at quantum theory on a nano level, through the smallest particles. But we have found that there are some phenomena on a macro level that work just like quanta. The resemblance is uncanny. Does that mean that we've discovered a new dimension? Maybe. Then again, maybe not.

3.6 The Expanding Space

All the stories, thought experiments, and logical conclusions you've read this far aim to show that observation is relative. They also depict how false our perception is regarding the results we ourselves have defined. Perception equals the experience of the effect—we don't experience the cause. We don't perceive the cause itself. Whatever we perceive does never prove reality, it only proves that we have received information coming from reality—so it means we have experienced the interactions. The information received is obsolete from the moment it is created. Past and future are both relative creatures. Whatever one considers past could be someone else's future, although, in reality, both of them originate from the *now*. If past and future can be perceived as present, we can question whether they are true.

We're on a journey toward understanding infinity, and the next stop is dissecting the expansion of the universe. We know that the whole world is expanding. But what does it really mean? It's a process that happens to space. It's inconsistent: at certain points, space expands faster than at others.

To understand the phenomenon, picture a fifteen-inches-long rubber band. If you take its two ends and stretch it, the two ends start to depart from each other very quickly. However, expansion slows down toward the center and becomes almost unnoticeable in the center itself. What happens to the universe is no different. The rim of the universe—the part that is still visible—is drifting away almost with the speed of light. Based on the distance between the sun and the earth, the receding is believed to be around 0.37 micrometer per second. This means that we're somewhere in the middle area of the imaginary rubber band. The expansion does not affect us now as much as it affects the still visible part of the universe, which, according to what we know today, is around fourteen billion light-years away from us.

The problem is that we don't know what exactly space is embedded in. If we did, we'd have the key to understanding why space keeps expanding at a growing pace. Based on our present knowledge, we suppose that space is surrounded by something that it is part of that other thing. Otherwise, it wouldn't be able to expand. Expansion in common terms means that space is forced out, and that's how the place of the material is created. If space is an unknown mass and it increases, then by definition, it must force something out somewhere in order to create the room it can occupy.

Space expansion confronts us with a complicated problem. We may feel like fish in an aquarium. We are surrounded by an impenetrable wall; still, we know that there is another world beyond. A world we will never be able to experience. Our aquarium wall is the fourteen to sixteen billion light-years. Why is this the limit for observing stars and galaxies? Is that where the universe ends? No, it doesn't end there. It simply means that the light of very distant galaxies, say, seventy billion light-years away from us, has not reached us yet. It would take another fifty-six billion light-years to observe those seventy-billion-light-years distant galaxies. Astronomers can get behind their telescopes each night, hoping that the light of an object that has not been there the night before might just reach them. It must be fascinating to witness a celestial phenomenon or object appear out of nowhere. The passing of time must make astronomers more and more hopeful. Or maybe not. Here's an interesting statement to put things into perspective: the farther we look into the universe, the lower our chance gets to discover new galaxies, and after a certain distance, it is impossible to discover new orbs.

We keep using the terms expansion and speed. Let's clarify their meaning when they're applied to the universe. Expansion is the *continuously decreasing* waiting time between two Planck lengths that follow one another. (If the waiting time is zero, it may mean we have reached the status of entanglement.) Speed is the *equal amount* of waiting time between more Planck length distances that follow one another.

Regarding the earth and the sun, we also observe expansion, but not only the earth and the sun get further and further from each other, also all other planets in our solar system do. It's just like inflating a balloon that has some kind of design on it that covers the whole surface of the balloon. The pattern is always visible, but as the size of the balloon grows, so does the pattern.

Observable galaxies can't easily help us in the subject of expansion either. The conditions we observe are a couple of million or billion years old. And on top of that, there is acceleration. To be more exact, the speed that the galaxy expands with at its rim. It may be as fast as the speed of light, or maybe faster; we can't exclude this option. Theoretically, it's possible to break the speed limit of light; it is even possible that expansion happens with a multitude of the speed of light—just think back to the chapter about light. The limit set by the theory of special relativity concerns the velocity of physical objects. And since the nature of space is not materialistic, its speed can surpass the speed of light.

Here's a thought experiment that might just open our eyes to the magnitude of the expansion problem. Imagine an experimental spaceship whose only task is to send one impulse of light every second to a receiving unit on Earth. Don't picture just any ordinary spaceship: this one has the ability to accelerate to the speed of light in an instant. On the launch site, the impulse generator is turned on, the laser light flares up at every second, and the receiver records the signal every second. Then our spaceship is launched. It immediately accelerates to the speed of light; it travels 300,000 kilometers in a second and sends out the first impulse of light. The receiving unit on Earth records the impulse the next second—that is how much time light needs to arrive. The spaceship travels on, and after the next second, it sends the next impulse to the unit on Earth, but by this time, it is 600,000 kilometers away from the receiver, so it takes two seconds for the light to reach the unit. After 900,000 kilometers, the third signal has to travel for three seconds in order to reach the receiving unit. Each second spent in the traveling spaceship adds another second to the waiting time. When the spaceship has traveled for a year and sends out the signal, it can be expected to arrive two years after the launching of the spaceship: the spaceship itself has been traveling for a year, and we have to wait for an additional year to record the impulse of light.

If we want to look beyond the observable universe and see the orbs being born now—at a distance of 15–20 billion light-years—that are accelerating from us maybe a bit faster than the speed of light, then we need to wait approximately 15–20 billion years. We'll then see their 15–20-billion-year-old state. This is the time we need to see that the planet is being born. And don't forget, the speed at which these orbs are moving away from us will only increase with time. It's like we're hitting wall after wall: first, there was the problem of the huge distance, which enables us to see distant orbs in the state they used to be in billions of years ago. Now, we see that even if we had a special device that could detect with the speed of light, we would still be unable to see reality as it is. With every passing second, we drift further and further away from the edge of the universe. With this device, we would see the exact same thing every day because the speed of the expansion would still increase on the periphery, and that would be our maximum speed.

However, there's a silver lining. Supposing we found a kind of energy that's faster than light, and whose interactions are not of a materialistic nature. Supposing we found several of these energies: one that travels twice as fast as light, one that travels ten times faster, and so on. Then we would only need to find their source conditions, and we'd have an actual time machine. We'd be able to extract the events, one slice at a time, from the information of the fast-expanding space. Right now, all information about the universe is stored and traveling toward us with the speed of light. We only need to tap into it to see the past that we now think is the future. But it doesn't look like we're going to get hold of this time machine any time soon. Time is still working against us, and we have less and less chance of discovering new planets with every second that passes. The statement now can fully be understood: the further we look into the universe, the lower our chances get to discover new galaxies, and after a certain distance, it is impossible to discover new stars or planets.

I say we switch the viewpoint and take a look at how the inhabitants of those distant planets experience the expansion of the universe that happens with the speed of light. Think back on the rubber band. When we stretched it and said that the ends expanded very quickly, we didn't set a reference. The end of the rubber band expanded faster than the middle part. But if we look at the end of the rubber band, we'll find that expansion is somewhat slower around the last two and three inches than right at the very end.

This leads us to the conclusion that the speed of the expansion is relative. It depends on the points of measurement. If compared to the expansion in the middle, it's fast. If compared to the end of the rubber band, it's slow. So the inhabitants of a distant planet that is in the part of the universe that expands with the speed of light, are most likely dealing with the same illusion of time as we do. Their time isn't any faster or slower; one day is just as long as the other. It's only us, looking from the earth, who see them rushing by with the speed of light. But there's a theory that the illusion of time changes when we travel at the speed of light. Still, it doesn't seem true in their case—because they only travel at the speed of light from *our* perspective, and compared to *us*. Two distant planets that are right next to each other will be motionless compared to each other. Space is a very peculiar thing because it doesn't have any material characteristics. We don't even know the minimum speed of space expansion. We can't even pull our trump card, the Planck length, since it is of a material nature.

In a previous definition of infinity, we found the absence of time. Maybe space is defined by time, i.e., movement. But with this idea, we're right back in the material world, and observing space, we haven't yet found material qualities. And to draw conclusions on space based on materialistic results is not very wise, I think.

3.7 Visiting the Neighbors—Dimension Style

There's an unsettling thought about the universe expanding at the speed of light: it's the idea of time dilation. According to the theory of special relativity, it is the difference in the elapsed time as measured by two clocks from two different frames of reference. If indeed everything is relative in this world, and we see that a particular planet is traveling at the speed of light (either on its own accord or because of expansion), then that orb *must* behave exactly as a spacecraft under the same conditions, for example. That means that the time in the spacecraft should pass slower *compared to our time.* Take, for instance, an inhabitable planet from the edge of the universe. Compared to the earth, that particular planet is traveling at the speed of light because it's located in an area where space expands at the speed of light.

Suppose we had some sort of device to observe the inhabitants of that distant planet—who would also happen to be at around the same level of technological development as we are. We would perceive everything and everyone almost completely still. They would look frozen. If we observed them long enough, we would notice that in about a year, the creatures and the objects around them moved about half an inch. At this pace, it would take a lifetime to watch one of them grab and drink a glass of water. We could conclude our research and say that time goes slower over there, because the planet and its inhabitants are traveling at the speed of light. We'd definitely have a think about how anything can progress at all in such a tempo. Why, simply building a house would take millions of years!

And this is exactly what time as an illusion means. Just think about it. The universe is believed to be roughly fourteen billion years old. But there's this distant planet, which—because of the expansion of the universe—is traveling at the speed of light. The time we count as fourteen billion years must mean more like 140,000 billion years for the inhabitants of that planet. (Let's just roll with this hypothetical number for the sake of this example.) It seems much longer because time goes much slower over there, or rather, time slows down. Which means, if I want to compare the system of the planet at the edge of the universe with that of our solar system, we get contradictory results. There will be two truths. One claims that the universe is fourteen billion years old, while the other is convinced that it's 140,000 billion years old.

And what would the inhabitants of that distant planet see of life over here on Earth? They would witness a chaotic sight; everything would happen in a frenzy. Skyscrapers would be built in a matter of seconds, people and cars would rush by with unfathomable speed. The aliens would start doing calculations. Considering this mind-blowing speed and the fact that their planet is still a lot faster than the earth (they would be aware of the expansion of the universe and that they move from the center with the speed of light, whereas we don't), they would count how much we could have lived since the formation of the universe. They would get the result of five hundred million years. (Again a hypothetical number for the sake of the example.) This number would be based on our speed and their system. They would wonder how we reached such a high level of evolution in such a short time. And once more, the two truths would be contradictory. We both would want to prove that we were right, based on the other system. This could remind you of the chapter about time when we talked about the issue with dog years. Well, the inhabitants of distant planets also experience life in its fullness; it does not feel longer or shorter to them. I think this helps understanding the relativity of time.

Maybe it's time for the question of what would happen if the universe stopped expanding? Well, there's a theory that time is the product of space expansion. It says that once the universe stops expanding, space doesn't feature time anymore. And without time, we'd have infinity.

Let's search for infinity. There must be a root cause somewhere that induced it all. If we found that, we would hold the key to understanding the formation of the universe—and possibly the formation of the multiverse too. The root cause is bound to be found beyond our usual dimension. Otherwise, we wouldn't have multiple theories trying to explain the universe's birth. Let's have a look at those few statements that may be good starting points.

- Infinity lacks all aspects of time.
- In an environment that lacks time, past, future, and present exist at the same time.
- Time is a necessary prerequisite of movement.
- There is no time in the system of light.
- There is no time at the level of quantum entanglement.
- There is no time and no velocity at the level of a single Planck length. There's only position change.

These statements do not contradict themselves or each other. They all have one thing in common: the notion of time. And it is time, or the lack of thereof, that makes all the difference in each particular situation.

Our thought experiments push us toward the idea that infinity is a possible starting point. Neither past nor future exists per se, because everything happens at the same time. However, it takes time to send out information about existence. If we think of infinity as the container of all possible events, it makes it possible to originate finiteness from infinity. Finiteness—the *now*—comes into existence once we add velocity and therefore time to any of these possibilities contained in infinity.

Does this mean that whenever something happens to us, it was predestined? No, this is not a matter of fate. But each possible event exists in infinity along with all possible states, all possible cause-effect scenarios, without paradoxes. Infinity contains it all. Velocity and time are the factors that select the possible events, and that's how we experience them as reality.

Wandering in space-time undermines every single frame of reference and turns everything upside down. If there's no fixed point to hold on to, we can't make sense of past, present, and future. It confuses us because we desperately need something to lean on, something that would be a strong enough point of comparison. We only understand *higher* when we have already experienced the *lower*. We live in a world of dualities; everything has its antonym: up and down, nice and ugly, cold and warm, big and small, fast and slow, close and distant. They're our own constructions; we need duality because we can't make sense of ourselves without them. Why do we need to experience ourselves in our dimension—while the possibility is there to access every single bit of information at the same time, at the same place, through infinity? Why did biological evolution choose a different path? It's as if the universe was forced to follow some invisible pattern to get to know itself.

4. Summing Up the Ingredients

In the previous chapters, we've examined everything that we thought was elementary to life. We have found that components we considered objective are, in fact, subjective. What is more, if we get to know them on a deeper level, their existence proves to be an illusion. The analyzing and understanding of elementary components is made harder by the fact that we have a problem understanding our own decision making methods. In several cases we choose the safe way: stay still and call it progress. We are content knowing that we are right and condemn those who happen to question the system. We don't look reality in the eye. Our ego is strong and would much rather deal with safe illusions. We give up on getting to know our reality. God forbid we dig a little deeper and have unanswered questions intrude our safe haven! By now, we have taught each other that there can be only one truth. We appoint someone who states the truth, then we comfortably, unconditionally follow him. Everything that fits in the truth teller's picture is fine, and we agree with it. Whatever doesn't fit must be denied. This process is what we call learning, education, understanding. We think we're open-minded because we listen to people contradicting the appointed truth teller—and then we persuade them there can be no other truth but the truth teller's.

All the while, history teaches us about progress. It teaches us about exploration and how thinking that we know enough is only the beginning of the journey. The thought of creation and a creative power at the root of everything has repeatedly presented itself as a possibility. The goal of this book is to analyze and clarify instead of pushing you in the one and only given direction, so I propose that if you are a member of any kind of religious or spiritual organization, go and ask your leader about the reason and purpose of creation. Then judge the answer for yourself from the perspective of logic. Be just as stern and ruthless as you were with this book. If the answer you received is logical and free from any form of self-justification, then you have found the root you were looking for. This form of truth exists, in my opinion.

II.

THE REALITY OF COINCIDENCE

1. The Way to Understanding

Where? What? How? We basically spend our whole life answering these three questions. We don't even notice how deeply they are ingrained in us, weaned into our life, theories, studies, and science. *Where, what, how*—always in this order—defines how we work, do research, live, and plan. Just think about your morning. I'm sure you are able to reconstruct everything you did and all the plans you had with exact precision if you simply follow the steps of this holy trinity. For instance, my plan for this morning is to go to my office and work on interest calculations using PL/SQL and Delphi programming languages.

It's clear by the example above that analyzing any process only makes sense if we answer all three questions regarding the plans. If I don't know where I must go, where I must be, it's pretty sure that I won't be able to accomplish anything. If I do know where I'm to go, but I don't know what I'm supposed to do, then again, nothing will come out of it: even though I'm at the right place, there is nothing to do. And finally, if I know where and what I'm supposed to do, but lack the necessary conditions, the *how*—for example, the software programming environment to me as a developer—I will most probably waste my time.

If we happen to change the order of these questions or wouldn't answer all of them, and, for instance, we chose to answer only *how*, we might get something that works, but it definitely wouldn't be working the same way as we expected at the beginning. It's a spontaneous act that creates something spontaneous. Should we want to be successful in creating, we need to be able to describe the process along the lines of *where, what,* and *how.* Regardless of whether creation aims to produce a material object, a theory or product, an action or a living being that wasn't there before.

Science basically is the description of observation. Most of the time, interactions lead us to conclude that there was a process behind them. The result of any plan will be such an end product that answers the questions where, what, and how. Any other approach is doomed from the start. They are the key to any problem, strongly interconnected in a hierarchical structure. This formula can be applied to any theory, be that the Big Bang, evolution, or something else.

Research—i.e., the comparison of various analyses—always focuses on an already existing product. The product itself answers where it does what and how. It's practically impossible to give an accurate description of the product's real purpose and nature by looking at the product alone. We would need to see the product design too.

Here's an example. Some archaeologists find an interesting artifact on an ancient Egyptian dig. It's a wooden board with four wheels, in surprisingly good condition. Carbon dating traces it back to the second millennium BC. Let's see what is the unknown variable in this equation.

- *Where*: it's a fact. Geographic location and time in history is clear.

- *What*: it's fiction. Archaeologists try to guess what the object was used for by looking at *where* and *how*.

- *How*: it's fact. The wheels can be turned.

There is no precise answer to the question of *what*, because we don't have the technical specification of the find. Archaeologists try to settle the question *what it does* by coming up with a possible explanation that fits into the hierarchy, between *when* and *how*. Unfortunately, we can't draw any conclusions on the purpose of an object simply knowing *where* it was used. And similarly, knowing how the object works doesn't let us conclude on its purpose. Whatever answers the archaeologists come up with to question *what does it do*, without having the missing information, there's no way we can accept their answer as true. They can't draw logical conclusions unless the object resembles another artifact found by archaeologists somewhere else, of which they know what it is for. In such cases, we usually use our imagination to fill in the gaps, but all people have different imaginative powers, so the end result can't be reality. It would paint a picture of reality as we'd like to imagine it.

Have you ever seen an electron? Me neither. It's just like Mrs. Columbo, no one has ever seen the fictional detective's wife, but we think she must be there. We drew a logical conclusion that electrons exist, and we include them in our calculations. If we go back a few decades in the history of the electron and take a look at the statements made on how it works, we will find these ridiculous. Exactly as ridiculous as our own theories will look in a matter of 15–20 years.

All we do know about electrons is the cause of their interactions, whenever an interaction happens. But we don't know where this electron—which is orbiting a proton—takes its energy from, although in some cases this energy is sufficient for it to have been doing so since the formation of the universe. It is similarly true to string theory: we don't know where the energy comes from to make it move.

Neither do we know why the electron needs to circle, how it does so, and what this miraculous energy is that starts the action. Let's look at an electron from the perspective of the question trio:

- *Where*: it's fiction. We've never been able to observe its location; we only see the results of interactions. Only the location of the interaction's result can be observed.

- *What*: it's a fact. Electrons can enter into interaction.

- *How*: it's fiction. We haven't seen how electrons function, how they do their thing; neither can we prove it.

It's exactly as if I gave a box of matches to a four-year-old child, and saying matches light fire, I'd leave him alone. What would happen? The child might have some vague ideas about fire but wouldn't know how it works, so having figured out how to light matches. most likely he'd burn the house down. If he knew how they worked and where not to light them, this could have a happy ending.

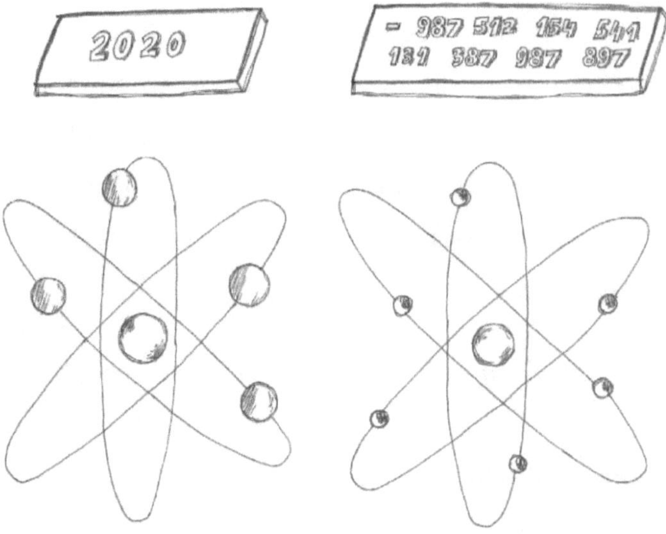

But let's not go on an atomic level; it surpasses our measuring capabilities. Let's stick to the micro-world instead and search somewhere closer, the closest we can. Let's examine ourselves, i.e., the human body based on the three key questions. Let's analyze the simplest case.

The human body contains somewhere around fifty trillion cells. Multiple cells create a unit called tissue, and tissues make up organs. What I would like to find out is how to convince cancer cells not to destroy the other cells that are useful for us.

As we see, this is such a level where with a well-functioning microscope, everything can be observed. However, we need to know what it is we eventually observe of the process. We need to pinpoint the exact moment when a cell decides whether or not it will divide abnormally. There's one requirement when it comes to decision-making: information. To make the cell make the right decision, we need to give it that kind of information. Since cells also communicate via information, it would be helpful to see how this communication process works.

- *Where*: it's fiction. There's no true proof what kinds of cells change. As of yet, we haven't been able to pinpoint the exact moment when a cell makes the decision to start behaving differently.

- *What*: it's a fact. Cancer cells attack healthy cells.

- *How*: it's fiction. We don't know how cells communicate with their environment. If we did, we would easily manipulate and mislead them. At present, we're still unable to do so; just think about antibiotics or learning cells. It's only the results of certain interactions that suggest some things about how cells work, but these results are not consistent. Neither do they give us proof of how communication units of cells work.

It's amazing that again, we only know one answer out of three: what the cells do. Never mind we have no idea how they work—we can't agree on what kind of cells are we talking about. And we developing medication based on this not-knowing that would affect the way cells function. We want to cause an effect on something of which we can't even describe the internal functioning. What could a pill do that we developed for something of which functioning we know nothing about? It's like we decided to take flip-flops with us to saw trees in a location of which the only thing we'd know was its name: The Forest. But there are thousands of forests, each filled with trees. How are we supposed to find the right place? It would take us forever to go through all of them. Wouldn't it be more efficient if we just had the GPS coordinates that would lead us straight to the tree we need to cut down? And even if we did find the Forest and the tree, what to do next? I'm not saying it's not fun sawing trees with flip-flops, but maybe that's not *how* trees are supposed to be chopped. Perhaps an ax or a chainsaw would be somewhat more useful.

We often think minor details are not important. We don't need to know all; it's the final outcome that we should be focused on. If cancer has a cure, fine. Following doctor's orders without questions is the way to go. Our world is spinning faster and faster; the web gets more and more tangled: we don't have the energy to deal with every single detail. For instance, do you think we'd ever drink milk if we knew where it came from?

Imagine going through the following checklist every time you wanted to get milk:

- Was the cow treated with antibiotics? And if so, are the residues still in the milk?
- Does the cow have any illness that hasn't been noticed and properly treated?
- What are the cow's living conditions?
- What is the cow's hormone level? Is it transferred to its milk?
- Is the cow descended from genetically modified cows?

And the list could go on. The point is, if any one single answer would be depressing, I wouldn't drink the milk at all. Maybe you wouldn't, either. It's simpler to suppose that any carton of milk is 100% natural and healthy. We accept things unconditionally when they come in a box. There's no need to have our own truth, right?

We've gone through a couple of examples, from Egypt to milk, following the three-question structure, but not one has passed the test. No matter how generous we want to be, we didn't see any signs that point at the possibility to reach accurate answers, either scientifically or technically. We take them as they come and don't question what's behind those results presented to us.

We software developers love Russian roulette. Maybe it has already occurred to you why your phone, your computer, and even your robot vacuum cleaner are constantly in need of software updates to fix bugs. Why aren't they right from the start? Fair question. That's why I teach my developers that a job poorly done requires just as much effort as a job well done—so let's do it well at the first attempt. But then why doesn't it work like that? Well, the problem is we don't know the details. We live in an age of recycling: we write new programs using old modules, classes, subroutines, and operating systems—it hardly ever happens that something is done from scratch.

Think about programming as playing with Lego. If we have the manual, we can see what each piece is for, as well as their sequence. By simply following the steps, we're able to make our building. This is the programming process. I'm sure you've played with Lego before, so you know what happens when you don't read the manual. If it's lost, we have absolutely no chance to build the exact thing on the box. The result may resemble it, but it's not up to our expectations by far. Say, we've got the manual, but we put one piece where it doesn't belong. It's possible that the building as a whole might not function properly because of that. And finally, if we have the instructions and follow them to the letter, but some pieces are missing, we may not be able to build what we were supposed to build. Not perfectly, that is.

Programming involves similar processes. If a colleague uses a module or class that was developed by someone else at some other time and it contains an error, it's enough to make the whole software defective. Maybe the programmer didn't read the class documentation thoroughly and he used it in a way that he wasn't supposed to. He didn't even notice that he used a module with an error or he used a correct one in the wrong way. But you as the user will notice — in a situation where the faulty program can prove its malfunction.

Programming languages are built on conditional statements (either true or false). There's always the chance that a truth value of a code changes because the condition itself changes. And again, there's the possibility that the programmer used a specific class for a different purpose than it was originally intended for. So you see, we produce impossible situations that make us harder to function, just because we don't care about the details. We simply use those program units which are supposed to be reliable, we wink at each other and hope for the best in silence.

But if that's the case, how come things do work?

Well — they don't. And if anything seems to be working, it happens at the cost of countless attempts of improvement and a lot of time and energy wasted — and it functions despite an unbelievably enormous amount of errors. None of our creations work because we understand their inner functions. Let me share a story with you. It shows that the world seems to be doing fine without us knowing the details. Once we start looking at the details, we realize that nothing works the way we expect. So, I was traveling in the car with my three-year-old son, listening to children's songs and nursery rhymes. We've been listening to all kinds of these for at least twenty minutes when I turned to my son, and asked him, "Sweetie, where do you think those boys and girls, and grown-ups are who do the singing?" My son pointed at the back of the car. I asked him, "How do you think they all fit in the trunk?"

My three-year-old understood that the sounds he heard were an illusion. He didn't cling to the idea that all those grown-ups and boys and girls were crammed into the trunk to sing for us. He terminated his faulty idea and didn't try to come up with other similarly unreasonable explanations, even though he didn't fully understand what was going on. (Of course I did try to explain).

As we dive into the chapters on coincidence, we'll need to keep an open eye and stand firm. Don't let yourself be influenced by me. Take everything with the right dose of doubt, lean on logic, and accept only statements that you cannot dispute by any logic.

1.1 Types of Coincidence

Elementary particles have several possibilities as to what happens to them. Whatever is possible to happen will eventually do so, if there are enough possibilities for that particular event to happen.

Conventionally, we distinguish between two types of coincidence: pseudo-coincidence and pure coincidence. Both types include the same three elements. The root cause, the event route, and the effect. Simply put, the starting event, the way to the consequence, and then the consequence. One of the trademarks of pseudo-coincidence is the fact that we know all three components. For instance, rolling the dice when we play board games is a case of pseudo-coincidence. We know that we cause it and that the result is achieved by the movement of the dice, as it rolls. We also know that the six-sided dice will inevitably show a number between one and six. We also know that if we throw it often enough, we'll encounter all six sides of the dice, and in equal proportion. This latter belief is based on the symmetry of the object. As it rolls, it will fall over its sides, edges, and corners a similar number of times. The reason why we believe it a coincidence is because the rolling is quick, therefore too many interactions take place in a very short time. That is, it rolls from one side to the other too fast for us. We do not see the exact event route, so we think the result is unpredictable. But it is possible to follow the movement of the rolling dice, and therefore, it's possible to reconstruct the path that leads to a given result. So, throwing the dice is not a real coincidence, but a pseudo one.

Pure coincidence is classified as such if the observer or the subject of the coincidence has no knowledge of the cause or the event route. Let's take a look at the following example, and see if we found real coincidence.

Imagine that it's fall and you're walking through a chestnut grove. Then a chestnut drops. You are just passing under the tree right at that moment, and the nut hits you right on the head. Then it bounces and lands on the ground. You probably think it's improbable. Out of all the places it could've fallen, did it have to hit your head? It's a simple enough story, and it has the answers. The route that connects cause and effect is clear: you were walking through a chestnut grove in fall, when chestnuts ripen.

Now let's just look at the same events from a slightly different angle. We grab an imaginary device, the size of a cell phone, with three buttons. The three buttons are fast motion, slow motion, and pause. Regardless of the rules of space-time, we can travel in our own time with this time-switching device.

Let's walk to the chestnut grove again and press pause. Now that there are no distractions, we go to the middle of the woods where there are lots and lots of trees. We set the device on fast motion, so that one day goes by as fast as one second. What will we see? An incredible amount of chestnuts will fall upon us during this one second; there's no way to avoid the shower.

If we survive the fall, let's give it another go. We wind time back to base point and hit pause again. This time, when we press slow motion, one second will pass with the speed of one day. Wherever we turn is safe. We can walk around and even stand under the chestnut trees; there are hardly any chestnuts falling, and if so, they are slower than dandelion fluff. Not one single nut will unexpectedly hit us on the head.

Well, we have seen that it's not a terrific idea to take a walk through a chestnut grove in the fall, but there's more. We may have realized that to experience sheer coincidence, time was necessary. If we slow down time—and with that, we slow down motion—there's no way that particular coincidence could ever happen. Analyzing our experiment with the chestnuts, we see that if we used slow motion and therefore had more time to observe how the nut falls from the tree, we could predict the place where it was going to hit the ground. We may even see the reason why the chestnut fell. Looking around, we perhaps noticed a big bird just landing on the branch where the chestnut was growing, and we realized it would fall because of the movement of the branch under the new weight.

One of the key conditions of coincidence is motion, and, consequently, time. And the faster the motion—i.e., the shorter the time between two Planck lengths—the less chance we have to interpret the outcome. Instead of precisely analyzing the effects of movements, which we deem impossible, we call it coincidence.

To define pseudo-coincidence very simply, it is related to too many things happening in a given period. To summarize what we've learned, we can state when we know the cause, the event route, and the effect, with the least probable events happening on the event route, the more certain we can be that the event is pseudo-coincidence.

1.2 Insurance and Coincidence

When you switch insurance companies because the new one offers a lower fee for your car insurance, they always ask the same question: "On average, how many miles do you drive per year?" You think they just want to know how skilled you are. But that's not the reason. These companies know that at your age, with your car make, and car color, there is bound to be an accident depending on the distance covered and a given time frame. They are curious to know these so that based on this information, they can work out the probability, and they can calculate the fee that would cover the financial loss when an accident occurs.

When someone recalled a car accident, they probably mentioned that it happened coincidentally. These so-called coincidences caused less or more damage. If you take all these "coincidences" and add personal information into the mix, you can calculate with exact precision how much money would cover the loss. If there really was such a thing as coincidence, it would be impossible to estimate the losses that occur from accidents, and there would be no way insurance companies could charge their fees. Insurance companies would simply not exist.

1.3 The Moral Side of Coincidence

In the following example, coincidence is literally a matter of life or death. There's an old, crumbling wooden bridge in a godforsaken part of the world, with a fifty-miles-deep chasm below. Falling in would be fatal. A few pieces of the deck's wooden planks are rotten. People cross the bridge every day for months and years on ends, and nothing happens. Then one day, there comes a heavily built man with his seven-year-old son on his shoulder. They weigh roughly twenty stones together, and this is how much pressure is applied to the deck with each step. One plank gives way. The man falls, the neighboring planks break too, and father and son fall into the deep. Both of them are instantly killed.

We can consider this to be a coincidence. This man was not the first heavily built one to step on the bridge, and not every plank of the deck was rotten. He simply took the wrong step. The plank didn't hold them. The police start investigating but conclude that the accident was nobody's fault. The father happened to step on a rotten plank. Everybody should calm down; father and son are the victims of coincidence. Nobody is responsible; their conscience is clear. But let's see what else happened in the background. Before father and son crossed the bridge, I walked across, too. I was half-way over the bridge when my wallet slipped from my hand, and all the change fell out and rolled away. So I went down on my knees to pick the loose coins, one by one. I noticed that one of the planks was almost completely broken; only the thin layer of paint held it together. After I picked up all the coins, I got up and walked on. I did see a bulky man coming from the other side with a child on his shoulders. The child could've been around seven or so, and he was half asleep. For a split second, it occurred to me that perhaps I should tell them not to step on the broken plank, but by the time I decided to open my mouth, they were already past me. I didn't want to shout after them; I just kept walking. What were the odds of the man stepping on that one broken piece anyway? A few seconds later, I heard a crack. I turned around, and I saw the man and his son falling into the deep.

If you don't know the whole story, you may think the accident was coincidence. But if you do, you could even accuse me of involuntary manslaughter, because I didn't prevent the tragedy. The cause was the same, the event route too, and the effect, tragically, was also the same in both cases. And that's how a coincidence turns out to be a pseudo-coincidence.

There's a highway in Europe that earned the 'most dangerous highway of the country' title in 2019. At that time, the road was being repaired, so the lanes were alternately closed for maintenance, which resulted in heavy traffic; sometimes, cars were stationary for an hour. Drivers became more and more frustrated because of the numerous regulations; everyone found the situation hopeless. Whoever had to drive through that section of the highway was sure to be delayed. There was no detour route, no alternative way. Regulations led to more and more offenses and therefore more and more accidents. Breaking the rules already leads to serious collisions within the first month of road maintenance. In the end, statistics showed that there was an increase in fatal accidents throughout the construction period. The conclusion was drawn that the number of fatalities could be estimated for every road maintenance situation.

If we knew in advance that construction work causes deaths because standard work and life conditions change, would we even start the job? I guess we would. At that point, a fatal accident would only be a distant possibility; no one would be responsible. We think we have no control over coincidental accidents. But is there really no one responsible? Isn't it a bit similar to the tragedy at the bridge? When analyzing the road accidents, we didn't find pure coincidence. All events were utterly predictable. In the case of a pseudo-coincidence, it's easy to modify the cause and therefore to predict how the effect changes.

But if all events we believe to be coincidence prove to be pseudo-coincidence, is there even such a thing as pure coincidence? We need to know it for the definition.

1.4 Coincidence in Statistics

A couple of years ago there was a TV program on mathematics on the Discovery Channel. They came up with an experiment to calculate the mean value. I still remember it and want to share it with you, hoping it will finally get us closer to what coincidence is.

Let's grab a two-pint mason jar, and fill it to three quarters with gummy bears. Don't count them, just pour. Then go out and ask two hundred random people to try and guess how many gummy bears there are in your jar. If someone's guess is clearly out-there (i.e., they mentioned a very small or a very big number on purpose, trying to sound funny), just dismiss it. Use the rest of the answers, and calculate the mean value. Then take the jar, and count the gummy bears. Compare the two results. You'll be amazed: the average guess will be a couple of gummy bears away from the exact number. If you asked two thousand people, the difference would most likely be only one or two gummy bears. And with even more guesses, you'd get the exact number of gummy bears.

Analyzing this repeatable experiment sheds light on the strange, yet controllable side of coincidence. Nobody actually knew how many gummy bears were in the jar; they only took a guess. Let's say that nobody happened to guess the exact number; still, there were several near misses. The mean value of the close estimates gave us the correct result. You didn't know how many gummy bears you poured in the jar; it really was a coincidence. And when people guessed, they also had a hope that they have the number right—by coincidence. So surely, this case must really be pure coincidence! But it's not. People always strive to give the best possible answer. And that best possible answer comes from everyone's own experience. When they guess higher or lower numbers, these results even out when calculating the mean value and ultimately lead to the correct result.

Maybe my assumption was wrong when I said that two incorrect values can never produce a correct result, since they just did. But it wasn't. In this particular case, what we have is not incorrect data in the sense that we are not trying to figure out a final number that we have no knowledge of. Quite the opposite: we actually know the exact result, but we don't know how the partial results lead up to it. Therefore, we can't come up with a rule, a logic to determine what kind of partial results we need. The gummy bear experiment is more of an analogy to the statement that a problem has thousands of possible solutions. And regarding pure coincidence, in this experiment we were able to pinpoint the exact reasons and effect, so we need to look further.

Based on what we know so far, I've made a chart. (We can only fill in the column of pseudo-coincidence since we haven't encountered the pure type yet.)

Feature	Applies to pseudo-coincidence	Applies to pure coincidence
The cause is known	yes	?
The route event is known	yes	?
The effect can be predicted	yes	?
The cause can be modified	yes	?
The effect can be modified through the cause	yes	?

All in all, we've looked at cases that looked as though they were a coincidence. Some of them were simple, others complex, and we even looked at extremely difficult cases, but we didn't find pure coincidence in any of them.

1.5 Coincidence and Time Travel

Get ready for an even more complicated thought experiment. I'll do my best to come up with an example of pure coincidence, and I can only do that by meddling with past and future. This story will be about Peter and George. Both of them have a car and drive to work every day. There's a certain intersection with a *give way to the right* rule that they cross each day. This means you have to give way to vehicles approaching the intersection from your right. Today, Peter is the one who happens to have priority to cross the intersection, and George must stop and give way.

In the morning, they both start their cars and take the same route to work as they always do. However, George has a lot on his plate; it's going to be a busy day for him at work, and his mind is already fixed on the problems he needs to solve. As both drivers approach the intersection, George is still lost in his thoughts. He fails to give way. Peter enters the intersection at that very moment. The cars crash. (Both of them slowed down, so the collision wasn't serious.)

This story must be considered a coincidence. The cars happened to reach the same place at the same time, and they simply crashed. We can see the event route clearly, and we can find the cause. But we can't definitely call it pure coincidence either. On the one hand, George must have reached that particular time and distance driven and other factors that inevitably lead to an accident, and on the other hand, it's not surprising to have accidents at intersections.

We should find a way to modify the event route to coincidence to avoid the accident altogether. Let's see what we could do to prevent the collision. Trying to turn this situation into pure coincidence, we find only one logical way. We must change the past based upon the knowledge of the future. As George steps out of his house, let's walk up to him and ask him what the time is. George will nicely tell us so, enabling two seconds to pass. And that's all we need. The short delay will change the event route; one side of the chain of events is modified. The two chains will be out of sync. There will be no accident, Peter will be gone from the intersection before George reaches it; he won't even see Peter's car.

George and Peter have just experienced a pure coincidence. And strangely enough, they have no idea that it was a coincidence that prevented their accident. They don't even know what the coincidental event was. They don't know that our short conversation with George ultimately saved them from colliding.

During this experiment, we seriously meddled with space-time. We changed an event we knew from the future, going back to the past. Thus we created pure coincidence. From Peter and George's viewpoint, it was a coincidence they know nothing about—there was no accident, they didn't even see one other that morning. It's only from our perspective that we can talk about some kind of coincidence. But because we were aware of the past when the accident did happen, and a present where there was no accident because we prevented it, it's definitely not pure coincidence. This is a clear case of a pseudo-coincidence—again.

In the chapter about 4+1 dimensions, we already talked about how the causes made in the past can be present in our now, and we can experience them as coincidence. But as long as the cause is known at least to the one who generates it, we're dealing with a pseudo-coincidence.

There are some new additions to the coincidence chart. And we need it because this chart will be more or less our only resort as we dive into the next chapter.

Feature	Applies to pseudo-coincidence	Applies to pure co-incidence
The cause is known	yes	?
The event route is known	yes	?
The effect can be predicted	yes	?
The cause can be modified	yes	?

The effect can be modified through the cause	yes	?
The cause can be modified from the future	yes	?
The cause can be modified from the past	yes	?
The effect may be relative	yes	?
The effect may be impossible to perceive, invisible	yes	?

1.6 Coincidence in Numbers

As I have mentioned in the very first chapter of this book, we can only generate pseudo-random numbers. They may seem they were random, but in reality, they are generated by a predictable algorithm. So, a more accurate definition of these so-called random sequences would be 'sequences that are hard to replicate'. One of the handiest random number generators is random.org. It uses atmospheric noise to generate random numbers. The program is based on the idea that it is impossible to reconstruct or replicate the number of events that are simultaneously happening in the atmosphere: electromagnetic waves coming from the sun, the signals transmitted by millions of satellites, the electromagnetic radiation of even more devices, and the cosmic background radiation. To picture the raffle, imagine a board spinning very quickly in front of you. All the numbers from 0 to 100 are written on it. Covering our eyes, we throw a dart at the spinning board. Whatever the dart hits will be our random number. Let's substitute atmospheric noise for the dart; photons and electrons will do the hitting, but instead of the spinning board, there will be a cycle that presents the numbers 0 to 100 at an extremely fast pace, incredibly many times. The number we call random is the one that is presented when the electron arrives. This site, random.org, is the go-to random number generator of famous casinos and online gambling platforms, and there are quite a few scientific experiments that make use of it, too. I also bought a couple of random numbers to do the following experiment.

Random number sequences are pushing the limits of our mental capabilities. When we see too many digits—let's say, a thousand of them—all at one place, we hardly know how to deal with them. We try to understand their value or look for a pattern of some underlying rule. But there's no way we could find anything at a single glance when we talk about a thousand numbers. If we wish to use algorithms to find a pattern, by definition, we must first know what exactly we're looking for. Then we could apply the appropriate algorithm. If we find it by chance, we are not one step closer to understanding the sample's randomness. There's no algorithm to calculate the complexity of a sequence, so our aim is to find out whether this seemingly random sequence carries any sort of information. If it does, the sequence wasn't truly random; there was a purpose behind it.

The first step of the experiment is to visualize the random numbers. We could simply begin with a mathematical algorithm, but these need the assumption that we know the process that leads us to the result. Since we don't even know what we're looking for, we're left with visualization.

It's simple enough: each random number will be placed on its own coordinate system. This particular experiment focuses on the random-like sequence of three numbers, 0, 1, and 2. They can only occupy one of the positions marked in gray.

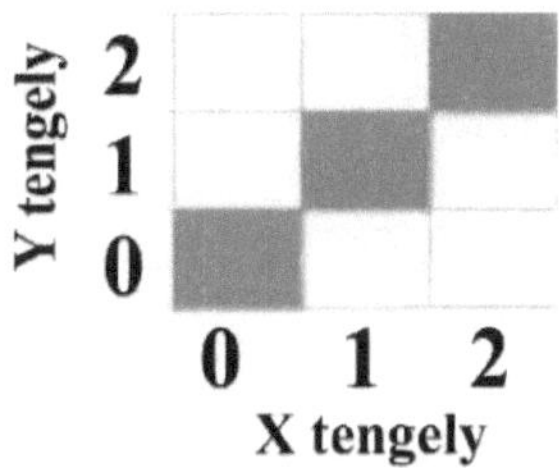

[Y axis; X axis]

The following figure shows numbers 0, 1, and 2 visualized.

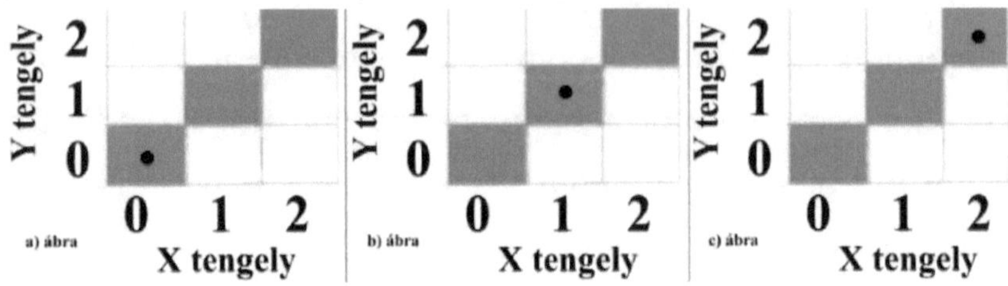

[Figure A; Figure B; Figure C]

Where there's a black dot in the gray square is our current number. Figure A represents 0, figure B represents 1, and figure C represents 2. If the random number is 1, then its coordinates on the X and Y axes are 1;1.

My 60,000 random numbers can have the value of either 0, 1, or 2. By looking at those 60,000 numbers, I could not spot any underlying pattern, so I put each of them on their separate coordinate system—I used 243 coordinate systems across and 243 coordinate systems down. We will analyze the numbers of the random sequence in this system. It's similar to the one below, only much bigger.

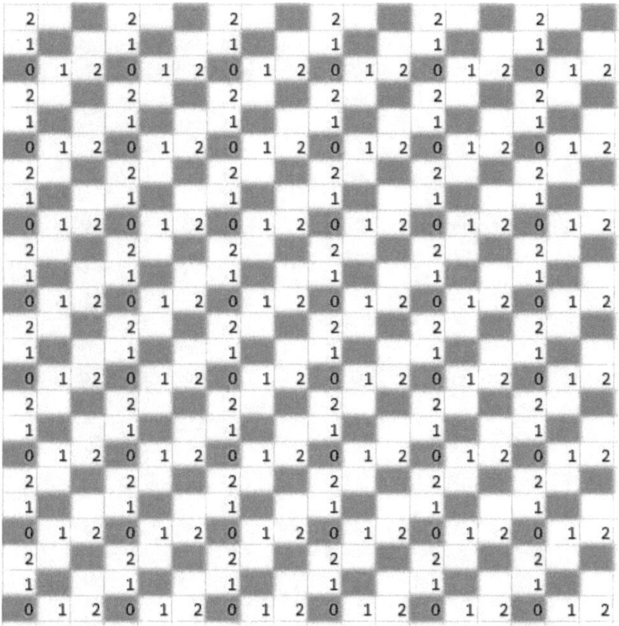

There are 59,049 random numbers in this 243 X 243 coordinate system. Each of them will be represented by one dot. To make up the complete picture of dots, I use a computer program I specifically wrote for this exercise. The first task of the program is to draw 59,049 invisible coordinate systems, and then it assigns each number of the random sequence to its designated coordinate system, putting the dots to draw the picture. Here is the image of this particular pattern of 59,049 random numbers.

It looks like the fuzzy screen of an old TV, while we were tuning the channel. This is not a coincidence; we used to see random numbers generated by atmospheric noise. In my picture, every dot represents the X and Y values of a given coordinate system.

Moving on to the next image, you'll see something quite intriguing. I have purchased an additional random sequence from random.org, in which the value of each number is either 0 or 1. My purpose was to find out what happens when two random sequences are mixed together. My first thought was that it should be another fuzzy, indiscernible image. There is one random sequence plus there's another. Still, the picture shows that they are two different random sequences mixed up.

At first glance, we don't see any serious differences. Let's reduce the size of the image because reduction always makes differences pop out. (To be able to discern the difference, I placed the original image resized underneath.)

The picture on the bottom is the original pattern made of 59,049 random numbers, reduced in size. But look at the picture on top: you see the mix-up in the matrix. I marked the problematic area with a black frame; you can see the unusual pattern. You can see that something happened to the random sequence. I know what it is; I did it. After the 20,000[th] number of the first random sequence, I inserted the 2,000 random numbers whose value was either 0 or 1.

We made the mixing of two random sequences visible. And we can even tell where it started by simply looking at the pattern. If we wanted to analyze the mixed pattern with mathematical algorithms, I'm not sure we'd discover that there are two sequences thrown in. I think we'd believe the pattern was caused by coincidence if we didn't know that it was done on purpose.

The next image also displays some sort of disruption in the pattern. It was caused by repeating the first hundred numbers five times, at five different spots after the 30,000[th] number. You can see that the forgery can be spotted. The image on top is the one that's been tampered with, and the bottom one is the original, that I didn't touch. In the top image, I marked the area in black, where the addition occurred. Try and see that those five sets with the same hundred numbers are identical to the first hundred numbers of the original sequence.

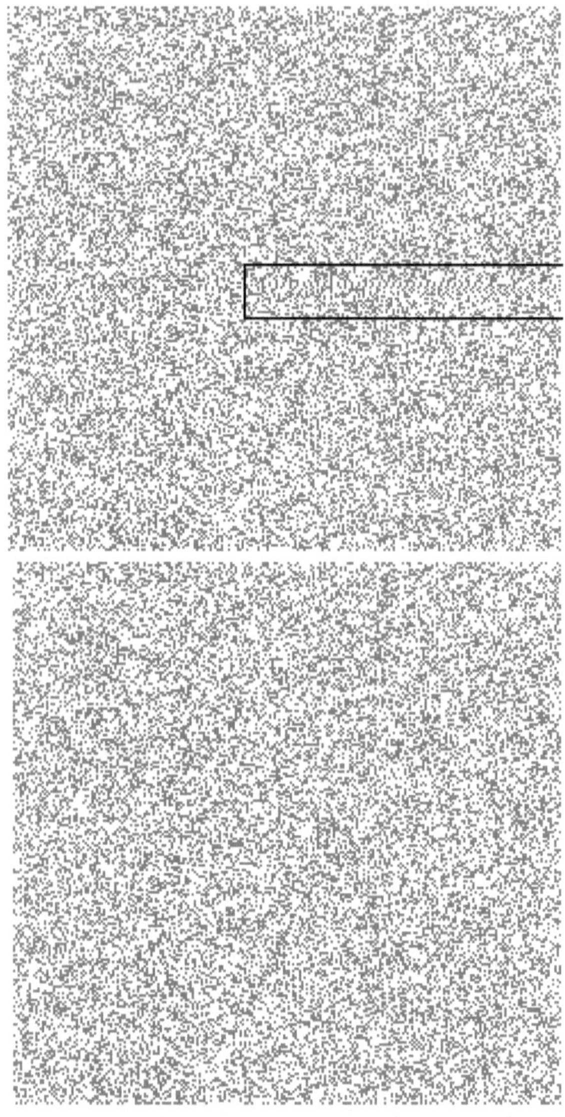

Let's do one more experiment. This time, the random sequence is made of 0s and 1s, each assigned to a 2 X 2 coordinate system. But we give it a twist: instead of another set from random.org, let's insert 3,408 binary numbers into the sequence after the 20,000[th] digit. These numbers come from a book's text, by having converted 426 characters.

Once again, the top picture shows the corrupted sequence, putting a black frame over the pattern modification, and the bottom image is the original one. In the corrupted sequence, there is a sentence that makes actual sense, and we found it.

These simple experiments show that random numbers can be manipulated, even of the best quality, and these manipulations can be made visible. Top-quality random sequences are generated according to a particular formula. If it weren't so, modifications couldn't be observed. Random sequences themselves already show patterns, although we may not see them. It doesn't matter; we understand anyway that if there was no pattern, modifications wouldn't be visible.

Should you be interested, I made several further experiments with another four million random numbers from random.org, and every now and then I spotted strange knots, disruptions in the images. They most likely came from atmospheric disturbances happening while the numbers were generated. So it's not entirely groundless to think that the random sequences generated by random.org can be influenced by manipulating the atmospheric noise.

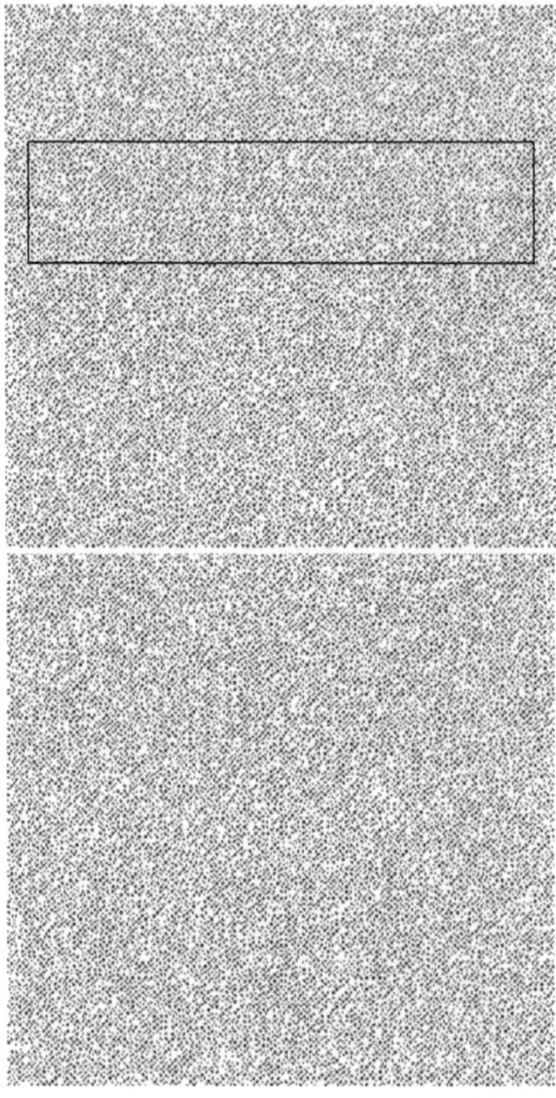

And maybe we can even go as far as to say that a random sequence can only be as random as we are blind to the connections. Once we figure out the generation algorithm, the sequence is not truly random anymore. We found that two different random number sequences are distinguishable—of course, we need to know what kind of random sequence we are looking for. It seems more and more true that the greater is our non-knowledge, the bigger coincidence we find even in random numbers. Random numbers appear to be truly random if we lack information on the way they were generated. Whoever knows all details of the process, will see them merely as a result of a complicated algorithm.

Therefore, random numbers belong to the category of pseudo-coincidence, too. Analyzing random numbers on a deeper level, through random sequences generated by random.org via atmospheric noise, our guiding principle should be chaos. We need to create an enormous chaos, as large as we can, so that with our current knowledge we wouldn't be able to model how we created it. Then, if we use a sample from this chaos, we do have random numbers. That would be a coincidence. Or wait, it'll be pseudo-coincidence because it originates from random.org, and we didn't find true coincidence here either. It would have been a splendid source of coincidence, though!

1.7 The Purpose of Random Numbers

We use random numbers as a means to prevent information from being understandable when getting into the wrong hands. In order to disguise information so that it is difficult to decipher, it needs to be converted into a particular pattern that is clear only to those who are aware of the conversion process. It is called encryption. Encryption is a pretty vast area; we are only concerned with its science in as far as it touches upon random numbers.

Encryption, code words, secret language—we may have used it in our childhood, for example, to message our friends in class when the teacher wasn't looking. Maybe we sent notes, with the words written backward. This is how the name Attila had a super-secret code version: Alitta. Once the teacher cracked the code, discovering that our encryption was so simple that we were just spelling words backward, he held the key, and every secret note we sent around in class became a public secret. Of course, we could come up with different ways of coding, for example swapping letters, but its fate played out similarly: once someone deciphered the code, all message content became readable.

The issue of encryption—i.e., protecting information—is probably as old as mankind. For thousands of years, people have had the same problems with encryption as me and my friends at school. A different method was needed. A system where the encryption process itself would depend on a constantly changing factor, so that the process would be completely unique and impossible to copy. It is important that this factor was constantly changing while the process remained the same, and that this changing created uniqueness.

Let me show you a very simple encryption method that shows the importance of coincidence. This method is safely built on and constant changes in the way that it creates its own key.

Let's assume two friends, Leo and Patrick, have to encrypt four words in a way so that the decryption needs both boys' keys. This means that Patrick will have one key and Leo will have the other. One key on its own is useless, but the combination of the two would give access to the protected information.

The first thing they do is create the encryption keys. That means that they convert the letters of the words into numbers. The information about which number stands for which letter will be known by a safe computer system only. The magic is changing the letters into numbers; each letter will have a unique number assigned to it. So when they want to send secret information, they only need to send the appropriate numbers. Of course, these numbers must be unique. If they are not, and by breaking their code we can also break their classmates Jack and Jill's code, it means the encryption wasn't a good one.

Leo and Patrick got these words to code: GIG, PIG, PIP, PI. We see they need to convert three letters (G, I, P) into numbers to have all words covered. In our key generator, the value of each letter is the sum of two numbers. We'll make a table out of them, asking the boys to roll a dice. Leo and Patrick take turns and assign one number to each letter. They must keep their numbers secret from each other; they store them in the computer.

Letter to be encoded	G	I	P
Patrick's secret number	1	4	5
Leo's secret number	3	2	6
Super-secret key	(1+3) = 4	(4+2) = 6	(5+6) = 11

In the table above we can see the super-secret key. Each letter has its own unique number. The boys rolled the dice to come up with them, so we can be almost sure that they are not exactly the same as someone else's super-secret key. The next step is to use the key and convert the words into codes.

Word	Coded version
GIG	4-6-4
PIG	11-6-4
PIP	11-6-11
PI	11-6

The secret code of the word PIG is 11-6-4, but this is only true within the system invented by Leo and Patrick, because they provided the numbers for encryption. Decryption is easy; we need to type the numbers into the computer, and the program will refer to the appropriate letter.

Coded word	11-6-4		
Secret key	11	6	4
Decoded letter	P	I	G

Patrick and Leo used the dice and miraculously, they didn't get identical numbers, so the experiment was successful. But this encryption method makes it clear that the variables should be unique and hard to guess. These numbers are impossible to create by people rolling dice, because all the dice and data recorders of the world wouldn't have sufficient capacity to come up with as many variables as we need.

These days, the role of the dice has been handed over to random code generators. But double-key encryption—albeit a lot more intricate—makes use of the same method. Both public and private encryption use coincidence to create the keys. Unfortunately, since creating random numbers needs time and motion, random numbers are only as reliable as the illusion of time.

1.8 Coincidence from Infinity

On our quest on coincidence, no matter where we looked, we kept running into time and motion. We know that time is relative. But if coincidence is tied to time through motion, then coincidence is also relative by definition. If infinity lacks time, then coincidence is practically impossible because it eliminates itself. In infinity every possibility, every eventuality exists simultaneously. It isn't viable for real coincidence to happen as all possibilities are already present. We couldn't find a single case of pure coincidence; we have all the more reason to believe that time is indeed absent from infinity. If it weren't so, we would've been able to find coincidence in finiteness—and an infinite number of it, on top of that.

After you have read my opinion, I believe you can guess that I think there is no such thing as coincidence, based on my experience. Whatever we think happened by chance shows a lack of information: we don't know the cause; we don't see the event route. And the less we know about them, the more coincidental we believe an event, a number sequence, or the past and the future.

2. Infinity, Creation, Physics

Let's try to recap what we've discovered this far, using the three key questions—*where, what, how*. Let's look at the origins of life and our universe.

Where did our universe possibly originate from? From infinity.

What does our universe do? It creates time, and consequently, finiteness.

How does our universe do that? Through velocity, it creates time, leaving infinity behind.

Answering the questions *where, what, how* always comes second, because at first, we need a cause: a stimulus that sets the chain of events in motion. Wherever we look, we see the effect or the event route—never the cause itself. No matter how we try, we can't find any single event, movement, or idea that has no cause. Maybe we don't know the cause there and then, maybe we never see the root cause, but only the cause of the previous event route.

By now we've seen that not being able to see something doesn't mean that it can't exist—it's rather a matter of perception: we're the ones who can't perceive it. And we now also know that perception is problematic in itself, because what we perceive is, in fact, some after-effect. It's not true by far that whatever we perceive is happening right now. (Looking at events closely, we see that it's never true.) The deeper we dive into finding the root cause, the more helpless the situation gets; we find every cause is, in fact, an effect. There's no logical way to find the root cause—except infinity. It's the only place where cause, effect, and event route do not depend upon each other, where they exist simultaneously. And in order for these sentences to make sense, this book needed to happen.

Analyzing infinity, we can ask the questions of what's the purpose and reason behind it? Why does it even exist? Well, with dualism at the chore of our worldview, it's not likely that we will ever fully grasp the meaning of infinity. It may be safe to say that infinity has always existed, since there's no time there. And as far as the purpose of infinity is concerned, there's only one logical perspective: the way infinity and finiteness are connected. Because infinity only makes sense when opposed to finiteness. And whether finiteness comes from infinity, or the other way round—instances of finiteness become possibilities in infinity—is really beside the point. These things only matter within our own dimension—if they matter at all. The opening lines of ancient texts about creation—oversimplified—go something like this: infinity became self-aware and felt the need for self-exploration. Since self-exploration is a process—and as such, it takes time—infinity created finiteness.

Following this train of thought, it all makes sense: infinity is exploring itself through finiteness, and it can keep doing so infinitely and for an infinite amount of time. Today's science and the lengthy, and at times complicated, logical assumptions all point to the very same conclusion about the origins of life.

Nevertheless, if you do find a case where any one feature does not apply to pseudo-coincidence—then you have discovered coincidence.

Let me show you the chart with the possible answers below.

Feature	Applies to pseudo-coincidence	Applies to pure coincidence
The cause is known	yes	no
The event route is known	yes	no
The effect is predictable	yes	no
The cause can be modified	yes	no
The effect can be modified through the cause	yes	no
The cause can be modified from the future	yes	no
The cause can be	yes	no

modified from the past		
The effect may be relative	yes	no
The effect may be impossible to perceive	yes	no

BEHIND COINCIDENCE

Afterword

We only get ourselves into trouble when we compare the truth of our own system with the truth of other systems. This sort of comparison leads to a lot of confusion and unreasonable results, contradicting ourselves. If you have questions about what it truly means to be or not to be—what the phrases *I am* and *I exist* stand for—then the book has done its job. All along, its purpose has been to shine a light on relativity, how our whole life is defined by double standards.

My goal with this book has been to unveil the notions of true and false carefully and make you see bit by bit that there is no absolute truth: everything is simply a matter of perspective. We have touched upon good and bad, realized they are indices in a system created by us ourselves; their validity is always limited to a certain context and can't be considered forever true. Therefore, being too quick to pass judgment does not only lead to some very awkward situations, but it is also the most certain path to humiliation—especially when this judgment is based upon 'absolute' truths.

If there wasn't a single thought in this book that you could identify with, then I must say: I couldn't be more thrilled. That's outstanding! It can only mean that you've got it covered: you don't need someone else to tell you what to think and how to think. You can stand on your own two feet and think—logically—for yourself. You don't need a truth teller. So this book only reinforced what you already knew: you are right.

If you are the kind of person who believes in truth-tellers, then again I consider my job done, because you've read the book. You gave me the possibility of convincing you that from now on you can think for yourself, you can make your own decisions, and you can grow on your own. We tend to agree and disagree without having a grounded reason. We are just stuck in our own system of reference, and we simply choose to go with the flow. It's very hard to leave our reference frame.

The Book That Happened gives you great possibilities to agree *or* disagree with what I wrote, since I strived to describe both options. Therefore, whichever option you chose, you gained. It's an unbelievably wonderful world we live in, and the only thing stopping us from seeing miracles is the people who say that we don't need to see and explore them. They believe it sufficient if we accept everything and ask no questions.

Acknowledgments

Kata, my wife—thank you for rolling with the idea of the book and for walking with me every step of the way. You understood that this book just had to come to life. And you did everything to stand by me—even when it meant carrying the weight of daily life and family on your own, for days and weeks and months on end. Thank you for stepping in, thank you for taking over my share. And thank you for our discussions—they left a mark in the book, that's for sure.

Krisz Nádasi, writer and this book's editor—the text would not have seen the light of day without your help. You treated my manuscript as you would treat your own. You left no stone unturned. Every possible question was asked, every single detail was explored, every misunderstanding was cleared. You weren't afraid to ask, to disagree, to argue—and ultimately, all the pieces fell into the right place. This was my first time writing a book, and like every other first time, this one seemed hopeless too. But your professional guidance and spot-on remarks helped me over the hump in no time. Thank you!

About the Author

The story of *The Book That Happened* goes all the way back to the 1980s. Ever since I was a child, I often pondered on these questions, but it wasn't until three years ago that I had the urge to put my thoughts into words and start writing. I'm not a writer, but I know what I'm talking about. I've had enough time to come up with solutions to millions of contradictions.

Professionally, I'm a lead software engineer. Even at work, I strive to live in tune with what I've written and create an environment where me and my colleagues can live our days and not just survive them. So every day is a new adventure to be explored to the max.

I'm never looking forward to the end of a day, no matter how frustrating and horrible it might have been—for could there be a more dreadful prospect than life getting shorter by a day?

Technicalities

Visit www.bookthathappened.com for any additional information and eventual modifications.

If you have any questions or remarks, just send an e-mail to attila@pergel.hu. I'm looking forward to hearing from you!

Why *The Book That Happened?*
–Behind the Scenes–

"If you can't put something into words, you can't achieve it either!" Do you agree? Most people don't.

Well, someone asked me this question some time ago, in the early 2000s. I couldn't answer right away, so I started thinking about it. The question and, later on, the answer became a part of my life: I agree. I can only achieve something if I'm able to put it into words—either write or talk about it. Many people still disagree. And I wonder why.

I can only consider myself an expert on a topic if I'm able to give a precise description of it: *where, what, how*—you know what I'm talking about. If I can give an outline of these three, I'm on the right path to an accurate description.

It's just like looking into the mirror and seeing reality. But what if there's a crack in the mirror? Then we only see something resembling reality. If we can't put our knowledge into words, then our knowledge is questionable. It's a sign that we're not yet sure what we mean.

And that's exactly the reason why I wrote this book: I wanted to make sure that I knew what I meant—that I really understood what I thought I understood. *The Book That Happened* had a forerunner: a 40-page-long manuscript that I first sent to an editor. The editor didn't like it. He didn't understand what it was about, or rather, didn't understand it the way I did. So it went back in the drawer, waiting to mature.

Then I gave it another try. But this time, I had the plan all laid out. I knew that I needed a partner. Someone who would be able to judge my thoughts and goals both objectively and subjectively. I needed an editor. An editor who is capable of unbiased judgment, who would treat my manuscript as their own, and still be able to keep the necessary distance.

Two editors have reviewed my draft—and rejected it straight away. The rejection was the biggest push in the process, because it made me realize that I did step on feet with this book. And I was more determined than ever. So I began researching self-development. I wanted to learn how to write a book. That's when I found Krisz

Nádasi's book with the very same title: *How to Write a Book* (in Hungarian: *Hogyan írjunk könyvet?*). It was exactly what I needed: it described my situation perfectly and answered my questions even before I realized I wanted to ask them. The book was brilliant. But at the same time, I realized that I would need a lot more time to fully grasp what it had to offer. And I always seemed to be short of time. So, as a last resort, I turned to the author of the book herself. I sent her the draft, and to my greatest surprise, she didn't send me packing but instead offered to meet with me.

I felt privileged. I've always loved learning from the best, and now I had the opportunity to have a private consultation with this brilliant author. Could things get any better? As it turned out, they could.

Krisz has a mentorship program, in which she works with both amateur and skilled writers, helping them write the book they've always dreamed of. I became part of this group, and the work began. I sent her the first part of the manuscript, and a few days later, I got her reply. She warned me not to be alarmed by the number of comments and remarks, because there would be a lot of them. I thought to myself, "Oh no, is it so bad that she's trying to prepare me for the worst?" and went ahead opening the document: it was indeed filled with her comments. But most of those comments were, in fact, questions. And then it dawned on me: Krisz wanted to make sure she fully understood what I had written. The questions were spot-on and in most cases, so deep, that I had to pay extreme attention and double-check my answers. I felt truly sorry for Krisz. Her job is one of the toughest: help writers do the best they can by constantly debating and arguing about every single detail—until the text becomes round and bulletproof.

Months have passed. Months filled with writing, replying, commenting, correcting. I believe the words 'good' and 'bad' are absent from Krisz's dictionary: things can be clear or incomprehensible, relevant to the subject or irrelevant—but never good or bad. She never said, "Don't do this or don't do that"; if anything, she said, "If you want to do that, do it this way!"

Finally the time came when Krisz declared the manuscript complete, and I also felt confident that I said everything I wanted to say. It was time to start editing. Now I know that editing is the most important step in the process: it's what makes the book come to life. Everything falls into place: paragraphs get switched around, words are replaced, sequences are put in order.

The book was finally ready. I've re-read and listened to it many times since then. I am more than convinced that the best decision was joining Krisz's mentorship program. I've learned how to write in a way to be understood, and I know what mistakes to avoid. I honestly believe that everyone should write at least one book in their lifetime. It's the most wonderful adventure of the mind. If you do decide to write, there's no one else I would more heartily recommend than Krisz Nádasi; she takes on English manuscripts as well (www.krisznadasiwrites.hu)!